Cucurbitaceae Crops - Breeding, Production, Current Research and Developments

Edited by Ertan Yildirim, Melek Ekinci, Murat Aydın and Esma Yiğider

Published in London, United Kingdom

Cucurbitaceae Crops - Breeding, Production, Current Research and Developments
http://dx.doi.org/10.5772/intechopen.1008643
Edited by Ertan Yildirim, Melek Ekinci, Murat Aydın and Esma Yiğider

Contributors
Andreea-Veronica Botezatu, Anna Busuioc, Ayse Gul Kasapoglu, Bianca Furdui, Devanshu Shukla, Ertan Yildirim, Esma Yigider, Esra Yaprak, John Wangai Kimenju, Kedar Mahadev Gheware, Krishna Kaushik, Kushal Chaturvedi, Manish Kumar, Melek Ekinci, Mohd. Wamiq, Mukesh Kumar, Murat Aydin, Rodica-Mihaela Dinică, Samuel Linguya Kimaru, Saurabh Tiwari, Sumeyra Ucar, William Maina Muiru

First published in London, United Kingdom, 2026 by IntechOpen
IntechOpen is the global imprint of INTECHOPEN LIMITED, registered in England and Wales, registration number: 11086078, 167-169 Great Portland Street, London, W1W 5PF, United Kingdom

For EU product safety concerns: IN TECH d.o.o., Prolaz Marije Krucifikse Kozulić 3, 51000 Rijeka, Croatia, info@intechopen.com or visit our website at intechopen.com.

British Library Cataloguing-in-Publication Data
A catalogue record for this book is available from the British Library

Cucurbitaceae Crops - Breeding, Production, Current Research and Developments
Edited by Ertan Yildirim, Melek Ekinci, Murat Aydın and Esma Yiğider
p. cm.

This title is part of the Agricultural Sciences Book Series, Volume 38
Topic: Agronomy and Horticulture
Series Editor: W. James Grichar
Topic Editors: Murat Helvaci and Olga Panfilova

Print ISBN 978-1-83635-179-5
Online ISBN 978-1-83635-178-8
eBook (PDF) ISBN 978-1-83635-210-5
ISSN 3029-052X

IntechOpen Book Series
Agricultural Sciences
Volume 38

Aims and Scope of the Series

The importance of agriculture cannot be overstated. It helps sustain life, as it gives us the food we need to survive and provides opportunities for economic well-being. Agriculture helps people prosper around the world and combines the creativity, imagination, and skill involved in planting crops and raising animals with modern production methods and new technologies. This series includes two main topics: Agronomy and Horticulture, and Animal Farming. This series will help readers better understand the intricacies of production agriculture and provide the new knowledge that is required to be successful. The success of a farmer in modern agriculture requires knowledge of events happening locally as well as globally that impact input decisions and ultimately determine net profit.

Meet the Series Editor

W. James Grichar has been employed with Texas A&M AgriLife Research for over 45 years with an emphasis on research in agronomy, plant pathology, and weed science. He obtained his BS from Texas A&M in 1972 and his Masters of Plant Protection in 1975. He has published 195 journal articles, over 330 research reports and briefs, 11 book chapters, and over 300 abstracts of profession meetings. He also directs research in many crops including corn, grain sorghum, peanuts, and sesame. He has held various positions in different professional societies including the American Peanut Research and Education Society, Southern Weed Science Society, and Texas Plant Protection Conference in addition to being Associate Editor for Peanut Science and Weed Technology. Significant accomplishments have included spearheading efforts to determine the optimum planting time for soybean production along the upper Texas Gulf Coast. These efforts have shown growers that soybean yields can be improved by 10 to 20% by following a late March to early April plant date. He also has been instrumental in developing a herbicide program for peanut production in the south Texas growing region. Through the development and use of herbicides that are effective against major weed problems in the south Texas region, peanut yields have increased by 25 to 30%.

Meet the Volume Editors

Ertan Yildirim is a Full Professor in the Department of Horticulture, Faculty of Agriculture, Atatürk University, Turkey. His research covers vegetable cultivation and breeding, greenhouse management, seed physiology, organic agriculture, and stress physiology. He was a visiting scientist at Cornell University in the USA and has served as Director of both the Vocational Training School and the Graduate School of Natural and Applied Sciences at Atatürk University. He is actively involved in national and international organizations. Professor Yildirim has released a registered garden cress cultivar and applied for the registration of bean and pinto bean cultivars. An author of over 250 publications, he frequently reviews journals and has participated in numerous international conferences, courses, and symposia.

Melek Ekinci is a Full Professor in the Department of Horticulture, Agriculture Faculty, Atatürk University, Türkiye. Her research focuses on teaching and researching topics such as vegetable growing, greenhouse management, seed germination and physiology, stress physiology, and mushroom cultivation. She has numerous publications to her credit and has attended numerous national and international courses, conferences, and symposia.

Murat Aydin is a faculty member in the Agricultural Biotechnology Department at Atatürk University and holds the title of Full Professor. His research focuses on plant biotechnology, molecular breeding, and abiotic stress tolerance in plants. He has played a key role in numerous national and international research projects and has published extensively in high-impact scientific journals. Additionally, he has participated in various national and international courses, congresses, and symposiums.

Esma Yigider has been an Assistant Professor in the Agricultural Biotechnology Department at Atatürk University. Her research focuses on plant biotechnology, abiotic stress tolerance, molecular breeding, and genome editing. Her work has been published in high-impact scientific journals, making significant contributions to the field of agricultural biotechnology. In addition to her research, Dr. Yigider has participated in various national and international courses, congresses, and symposiums, keeping up with the latest developments in plant biotechnology and molecular genetics.

Contents

Preface

Cucurbitaceous crops play a crucial role in agriculture, nutrition, and public health worldwide. Cultivated in diverse climates and cultures, they represent crops that are both of long-standing tradition and modern relevance. Although widely grown and consumed, their improvement through breeding and production strategies is far from straightforward, as it is influenced by their evolutionary diversity, distinctive morphological traits, and complex physiological and molecular processes. At first glance, cucurbits appear as everyday vegetables on our tables. However, a closer look reveals remarkable diversity in their genetic makeup, bioactive compounds, and adaptive abilities, which resist simple explanation. Each species responds uniquely to ecological conditions, cultivation practices, and storage or processing methods. For this reason, a thorough understanding of their biology and utilization requires knowledge that spans breeding science, biotechnology, crop production, and postharvest management.

In recent years, research into cucurbits has expanded considerably. Breeding and genetic studies have created new opportunities for biofortification and resilience under climate stress. Equally, advances in in vitro culture, such as micropropagation, organogenesis, and somatic embryogenesis, have accelerated plant regeneration studies. At the same time, investigations into phytochemical profiles and bioactive properties have highlighted the value of cucurbits not only as sources of food but also as contributors to traditional and modern medicine.

This book was prepared to bring together the latest knowledge on cucurbit crops. The chapters explore breeding and evolutionary perspectives, cultural practices, postharvest applications, biotechnological tools, and the role of indigenous cucurbits in food and nutrition security. By presenting these aspects collectively, the book provides a comprehensive view of the scientific progress and practical significance of this plant family.

Ertan Yildirim, Melek Ekinci, Murat Aydin and Esma Yigider
Faculty of Agriculture,
Atatürk University,
Erzurum, Türkiye

Chapter 1

Introductory Chapter: The Scientific and Agricultural Importance of Cucurbitaceae

Ertan Yildirim

1. Introduction

The Cucurbitaceae family is among the most economically, nutritionally, and scientifically important plant families globally because it accounts for a large share of agriculture. Cucurbits, including cucumber, melon, watermelon, squash, pumpkin, and bottle gourd, are among the most widely cultivated vegetable and fruit crops in the world, with around 118 genera and almost 1000 species [1]. Their high water, vitamins, antioxidants, and dietary fiber content make these crops one of the staple parts of the human diet in many places. Besides their nutritional importance, cucurbit crops play an important role in the global agricultural economy with fresh eating, food processing, seed production, and international trade.

These crops are second to none. They are deeply embedded in the food systems of pan-continental co-existing societies, providing not just human nutrition but also economic support for rural populations, agro-industrial corridors, and international trade routes. They are grown in temperate, subtropical, and tropical climates and exhibit remarkable adaptation to diverse agro-ecological zones. Its wide genetic variability has allowed it to adapt to numerous environmental conditions, from dry and semi-dry zones to humid production systems and greenhouses. Furthermore, the increasing global demand for fresh vegetables, functional foods, and value-added horticultural products has added strategic significance to cucurbit crops in sustainable agriculture [2]. Cucurbits cover millions of hectares of cultivated land worldwide. The ones with the greatest production volume among them are watermelon, cucumber, melon, pumpkin, and squash. These crops are grown in open-field agriculture, non-heated cultivation systems, and heated production in greenhouses. Aside from their economic importance, cucurbits have become valuable plant model systems. Cucurbit species are commonly used as experimental models in studies addressing sex determination, vascular biology, fruit development, and plant stress physiology. The reproductive biology of many cucurbits has an especially promiscuous history in plant experiments, providing distinguishing information about the regulation of plant developmental processes and hormone function at various levels, situated between species and organelles on or around a paradigm herbaceous vine.

Cucurbits are not only economically relevant but they also present remarkable biological interest. They have distinctive morphological traits, including indeterminate growth, climbing or creeping growth habits, tendrils (twining stem modifications), monoecious or dioecious flowering patterns, and specialized fruit types called pepos. Being a model species for floral development and plant hormone regulation, their reproductive biology, sex expression mechanisms, and pollination systems have been extensively studied. Additionally, multiple cucurbit species have been utilized as experimental systems in studies related to physiology, genetics, and molecular biology [3].

Recent decades witnessed the advent of rapidly advancing plant sciences, which have dramatically reshaped cucurbit research. New frontiers have been opened in crop improvement by developments in molecular genetics, genomics, biotechnology, and precision agriculture. Next-generation solutions in genome sequencing projects, high-throughput phenotyping to identify traits in the field more rapidly, and molecular marker technologies and gene-editing tools, including CRISPR/Cas systems, are reinventing breeding strategies [4]. This leads to new cultivars with applications including higher yield, better nutritional value, and greater resistance against biotic and abiotic stress. Meanwhile, challenges to cucurbit production are increasing. Climate change, defined by increasing temperatures, irregular precipitation regimes, salinity, and extreme weather events, presents severe threats to yield stability and crop quality. Emerging diseases and pests, increasing adaptability, and pathogen races indicate a need for innovative integrated sustainable management. Challenges, including post-harvest losses, inefficiencies throughout the supply chain, and quality degradation in storage and transport processes, compound production systems. Meeting these complex challenges requires a holistic, interdisciplinary view.

Cucurbits are explored for ecology within plant-environment interactions, pollination biology, seed dispersal mechanisms, and ecological adaptations. We assess the effects of climate change as it progresses globally on cucurbit production systems, highlighting demand for resilient genotypes and climate-smart [5]. Physiological characteristics of the cucurbit plant distinguish it from other vegetable crops. They show rapid vegetative growth and fruit development, which depend on photosynthesis in source leaves and on the transport of assimilates. One of the most prominent features of cucurbits is their large vascular system to facilitate the rapid transport of photosynthetic products from source leaves to developing fruits and vice versa. This effective carbohydrate partitioning is critical in developing substantial fruits such as watermelon and pumpkin.

Cucurbit plants exhibit intricate hormonal regulation of growth and reproductive development. Phytohormones ethylene, gibberellins, and auxins are important regulators of floral sex expression, fruit set, and fruit enlargement. Hormonal pathways are additionally modulated by environmental factors, such as temperature, photoperiod, and nutrient availability, to regulate these processes. Countless accessions of Cucurbitaceae, a genetic resource that needs to be unearthed and used in crop improvement, speed farming. Wild species are often a source of genes for resistance to diseases, pests, and environmental stresses in cultivated cucurbits. Plant breeding programs have produced many improved cultivars that exhibit higher yield, fruit quality, disease resistance, and adaptability to the environment. Traditional breeding methods have also been supplemented with modern molecular tools, including marker-assisted selection, genomic mapping, and transcriptomic analysis.

Expandability of cucurbit crops is also being developed by biotechnological methods, including gene editing and functional genomics [6]. These resources help researchers understand complex traits and develop cultivars that can maintain productivity in a changing environment. Cucurbit crops are among the world's most popular and beneficial vegetables, but face many production challenges. These are pests and disease susceptibility, environmental stress sensitivity, and climate change impacts on crop productivity. Many regions are affiliated with increasing temperature, erratic rainfall distribution, or soil salinization, all of which put cucurbit culture under jeopardy. There is no single way to tackle these challenges, and integrated approaches of plant breeding, better cultivation techniques, and cutting-edge molecular approaches will be required.

2. Future perspectives and research directions in cucurbit science

Cucurbitaceae will continue to be critical for the region's economy and agricultural ecological conditions, providing more nutrients and significant benefits. However, they face growing pressures from environmental degradation, shifting consumer trends, and a demand for sustainable farming methods that warrant innovative new research strategies. The contribution of future advances in cucurbit science will depend heavily on the merging of modern technologies with traditional horticultural knowledge to improve productivity, adaptation potential, and fruit quality. Among them, one of the most promising areas is the application of genomic and molecular-based tools in cucurbit breeding programs. As whole-genome sequences are becoming available for major cucurbit species, new opportunities arise to identify genes underlying crucial agronomic traits related to yield, disease resistance, and stress tolerance, as well as fruit quality. Genomic selection, marker-assisted selections, and gene-editing technologies (e.g., CRISPR/Cas systems) will further accelerate the development of improved cultivars. These strategies will give breeders the opportunity to specifically target favorable traits while significantly streamlining the process of cultivar development.

3. Conclusion

The Cucurbitaceae family is among the most important groups of crops providing food security, nutrition, and agricultural economy across the globe. In addition to their significance as food crops, cucurbits are important model systems for plant physiology, reproductive biology, and stress responses. Going forward, there must be continuous study of cucurbit biology, genetics, and cultivation practices to meet progress on future agricultural challenges. The combination of traditional knowledge with modern science techniques will enhance the development of high-quality cucurbits as well as sustainable agricultural systems that are a necessity to ensure that vegetable crop production is appealing for an ever-increasing global population.

AI tools disclosure

The author acknowledges the use of Grammarly for language editing.

Author details

Ertan Yildirim
Department of Horticulture, Faculty of Agriculture, Atatürk University, Erzurum, Turkey

*Address all correspondence to: ertanyil@atauni.edu.tr

References

[1] Weng Y, Sun Z. Major cucurbit crops. In: Wang Y-H, Behera TK, Kole C, editors. Genetics, Genomics and Breeding of Cucurbits. New Hampshire, USA: Science Publishers; 2012. pp. 1-16

[2] Ebert AW. The role of vegetable genetic resources in nutrition security and vegetable breeding. Plants. 2020;**9**(6):736

[3] Grumet R, McCreight JD, McGregor C, Weng Y, Mazourek M, Reitsma K, et al. Genetic resources and vulnerabilities of major cucurbit crops. Genes. 2021;**12**(8):1222

[4] Ahmed N, Zhang B, Deng L, Bozdar B, Li J, Chachar S, et al. Advancing horizons in vegetable cultivation: A journey from ageold practices to high-tech greenhouse cultivation—A review. Frontiers in Plant Science. 2024;**15**:1357153. DOI: 10.3389/fpls.2024.1357153

[5] Knapp J, Osborne J. Cucurbits as a model system for crop pollination management. Journal of Pollination Ecology. 2019;**25**(9):89-102

[6] Feng J, Wang N, Li Y, Wang H, Zhang W, Wang H, et al. Recent progress in genetic transformation and gene editing technology in cucurbit crops. Agronomy. 2023;**13**(3):755

Chapter 2

Chayote: An Important Cucurbitaceous Vegetable Crop

Krishna Kaushik, Kedar Mahadev Gheware, Mukesh Kumar, Mohd. Wamiq, Manish Kumar, Kushal Chaturvedi, Devanshu Shukla and Saurabh Tiwari

Abstract

Sechium edule (chayote) is a nutrient-dense cucurbit with notable agronomic, nutritional, and medicinal importance. Native to Mesoamerica and cultivated across tropical and subtropical regions, it is recognized for its vigorous growth, morphological diversity, and high-value phytochemicals. This chapter provides an integrated overview of chayote's taxonomy, botany, pollination biology, and global distribution, emphasizing its cultural and historical significance. The species offers health benefits through its edible parts—fruit, seed, stem, and root—due to its rich content of dietary fiber, vitamins, flavonoids, phenolics, and saponins. Pharmacologically, *S. edule* exhibits antioxidant, antihypertensive, antidiabetic, and anticancer properties. Despite its potential, it remains underutilized due to breeding limitations, the absence of standardized cultivars, and insufficient germplasm conservation. The chapter discusses recent advances in breeding, molecular markers, and *in vitro* techniques for genetic improvement and conservation. Cultivation practices, pest and disease challenges, and postharvest management are reviewed, alongside key constraints such as biotic stress, limited genetic resource use, and market barriers. Strategic recommendations include integrated conservation, enhanced genetic research, and market diversification to support its role in sustainable agriculture.

Keywords: chayote, nutritional value, medicinal properties, genetic improvement, sustainability cultivation

1. Introduction

Chayote is known as a vegetable pear or Swartz in many parts of the world. It belongs to the Cucurbitaceae family. Chayote is cultivated worldwide [1] due to its high nutritional [2, 3] and bio-functional value [4]. Brazil is the largest producer of chayote, followed by Costa Rica, the Dominican Republic, and Mexico [5]. Chayote is commercially cultivated in India in Tamil Nadu, Karnataka, West Bengal, Uttarakhand, Himachal Pradesh, and the North Eastern Hill Regions [6]. According to Refs. [7–10], in different regions of the world, chayote is known by different names such as Chayote in Mexico, Christophere, mirliton, vegetable pear, mirliton choko in Australia and New Zealand [11], Starpreciante, Citrayota, Citrayote in Ecuador and Colombia, Chuchu in Brazil, Machucha, Caiota, Piniela in Portugal, Chow-Chow in India, Cho-Cho in Jamaica, Sayote in the Philippines, Guseril in Guatemala and El-Salvador, and Colbaza Pear or Iskus or Pear squash in Nepal.

It is spotted as an herbaceous perennial climbing characteristic plant [12, 13] with vine-like structure, having tendrils for climbing, with pale green to white in color, mild in flavor, crunchy in texture, with single-seeded fruits that have tuberous roots [12, 14]. It can be used in the cosmetic, food, and pharmaceutical industries due to its richness in phenolic compounds, polysaccharides, cucurbitacins, proteins, and vitamins [15]. Several studies say that the chayote is usually consumed as a fruit, and the peel or leaves are rarely consumed [5, 9, 10, 12, 16–18].

2. Botanical classification and morphology

2.1 Taxonomy

2.1.1 History of taxonomy

The plant of chayote was first documented by P. Browne in 1756 in "Civil and Natural History of Jamaica". Initially placed in various genera before settling into its current classification within the genus *Sechium*. The species name "*edule*" reflects its edible nature. In 1763, Jacquin and Adanson independently classified the plant, with Jacquin placing it in Sicyos and Adanson in Chocho. In 1788, Jacquin moved the plant to his newly created genus Chayota. After 12 years in 1800, Swartz formally placed it in the genus *Sechium*, using the name *S. edule* (Jacq.) Swartz, which is still used today (**Table 1**). Taxonomically, chayote has 10 species of which eight are wild *S. chinantlense, S. compositum, S. hintonii, S talamancense, S. panamense, S. pittieri, S. venosum,* and *S. vilosum* from Mexico to Panama, and only two are commercially cultivated *S. tabaco,* only cultivated in Costa Rica, and *S. edule,* widely cultivated throughout the America and other regions [19, 20].

A wide variation in the *S. edule* chromosome number has been documented in the literature. Some studies agree that the haploid and diploid numbers of this species are n = 12 and 2n = 24, respectively [21, 22], while others report accounts of n = 13 and 2n = 26 [23], 2n = 28 [24], or 2n = 22 [25].

Kingdom	Plantae (Plants)	[11]
Phylum	Magnoliophyta (Flowering plants)	
Class	Magnoliopsida (Dicotyledons)	
Order	Violales (a diverse order of flowering plants)	
Family	Cucurbitaceae (Gourd family)	
Genus	*Sechium*	
Species	*Sechium edule* Jacq Sw.	

Table 1.
Taxonomic classification.

2.2 Botanical description

Chayote (*Sechium edule*) is a fast-growing, perennial, and monoecious climbing vine of the Cucurbitaceae family. It originates from a thickened primary root system that produces adventitious tuberous roots. The angular, grooved, and glabrous stems often emerge in multiples from a single root in cultivated forms. Basal stem regions are lignified, whereas the apical areas develop numerous slender, herbaceous branches. Leaves are ovate-cordate to suborbicular, glabrous, and borne on grooved petioles (8–15 cm). Leaf blades (10–30 cm) exhibit 3–5 shallow lobes with cordate bases and pointed apices. Tendrils, typically 3–5 branched, provide structural support.

Chayote is monoecious with unisexual flowers. Male (staminate) flowers form in erect, pedunculate racemes (10–30 cm), particularly in wild types. These have short puberulent pedicels and a glabrous receptacle. Flowers possess five reflexed sepals (4 × 1 mm) and five triangular petals (6–7 × 2–3 mm), green to greenish-white. The androecium consists of five fused stamens forming a central column with free apices bearing oblong anthers; anther configuration includes bithecous and monothecous forms. Female (pistillate) flowers are solitary and located in the same axils as male racemes.

Fruits are borne singly or in pairs and vary widely in shape, size, and color. They are fleshy or fibrous with prominent ridges, furrows, or lenticels. The mesocarp ranges from pale green to whitish, with taste varying from bitter (wild types) to mild or sweet (cultivars). Each fruit contains a single large, ovoid, and smooth seed. Cultivated types often exhibit vivipary, while wild types typically germinate only after fruit detachment.

2.3 Pollination ecology of chayote

Pollination in *Sechium edule* is predominantly entomophilous, involving a diverse array of insect species. Studies indicate no significant variation in fruit set between plants subjected to open pollination and those subjected to self- or cross-pollination, suggesting a broad compatibility and efficient pollination mechanism [26, 27]. Furthermore, fruit development does not appear to be influenced by the quantity or frequency of pollen deposition on the stigma. However, experimental cultivation under greenhouse conditions devoid of pollinating insects has demonstrated that in the absence of biotic pollinators, fruit development is impaired, resulting in premature abscission of immature fruits [28]. These findings confirm the essential role of insect-mediated pollination in successful fruit development and reproductive success in chayote.

The dependency of chayote on insect pollinators may partly explain its extensive distribution and adaptability; however, this reliance presents a major constraint for the conservation of genetically pure lines. This poses challenges not only for commercial cultivation and traditional agricultural systems but also for germplasm conservation efforts in gene banks. The effectiveness and diversity of pollinators are influenced by several ecogeographical and anthropogenic factors, including altitude, latitude, and pesticide usage [24, 26, 28]. Notably, species of the genus *Trigona*, which are recognized as highly effective pollinators of chayote, are predominantly found in medium- to high-altitude regions that are generally free from pesticide applications. In contrast, *Apis mellifera*, another key pollinator, is more commonly associated with lowland commercial plantations where chemical inputs are frequently employed.

Additional secondary pollinators include wasps from the genera *Polybia*, *Synoeca*, and *Parachartergus*, as well as smaller *Trigona* species. These variations in pollinator communities highlight the intricate interactions between environmental conditions and pollination biology in chayote cultivation.

2.4 Origin and distribution

There is currently no direct archaeological evidence to determine the antiquity of *Sechium edule* (chayote) cultivation. The lack of physical remains such as pollen grains, macrofossils, or desiccated fruit structures in archaeological contexts is likely due to the fruit's fleshy nature and the presence of a single soft-seeded testa, which is poorly suited to long-term preservation. Consequently, the reconstruction of the domestication history and geographical origin of *S. edule* has relied heavily on ethnohistorical records, artistic representations, linguistic data, and analyses of the ecogeographic distribution of genetic diversity among wild and cultivated taxa.

Ethnohistorical records indicate that *S. edule* was cultivated in pre-Columbian times, particularly in Mesoamerica. One of the earliest descriptions is attributed to Francisco Hernández during his visit to Mexico between 1550 and 1560 [29]. However, the introduction of the species to the southern regions of the Americas appears to have occurred only after the Spanish conquest [26, 30]. Linguistic evidence further supports a Mesoamerican origin, as the common names for chayote across Latin America—particularly the term "chayote," derived from the Nahuatl—remain consistent, with only minor regional modifications. These names are also retained in other global regions where the crop has since been introduced. Additionally, pre-Columbian pottery from Mexico and Central America provides artistic representations of the fruit, offering visual confirmation of its cultural and agricultural relevance in antiquity [30].

The most compelling evidence for the origin of *S. edule* lies in its ecogeographic distribution. Numerous botanical explorations and taxonomic surveys [31–36] have consistently reported that the highest diversity of cultivated forms is concentrated in southern Mexico, Guatemala, and Costa Rica, primarily at altitudinal ranges between 500 and 1500 meters above sea level. These findings align with patterns observed for crop centers of origin, where maximum genetic variation typically coincides with domestication zones.

The distribution of wild relatives of chayote further corroborates this hypothesis. Morphologically similar wild taxa of *Sechium* occur within the same altitudinal and geographical parameters as cultivated types, particularly in southern Mexico. Notably, the so-called "wild types" of *S. edule*, previously misidentified as feral forms, have been documented in the states of Veracruz, Puebla, Hidalgo, and Oaxaca [26, 36, 37]. These populations are predominantly located in humid environments—such as ravines, riverbanks, and waterfalls—within montane rainforest ecosystems, as well as in the lower transitional zones of evergreen or semievergreen seasonal forests, at elevations ranging from 500 to 1700 m above sea level.

In addition, a newly described endemic species from northern Oaxaca—initially considered a wild form of *S. edule*—was later distinguished taxonomically as a separate species [38], highlighting the complexity and richness of wild *Sechium* diversity in the region.

Reports of wild populations of *S. edule* from Java, Réunion [39, 40], and parts of Venezuela [41] remain unverified due to the lack of herbarium specimens or genetic studies. It is plausible that these populations may represent feral escapes from cultivation, rather than genuinely wild progenitors, as suggested by expert communications [42].

3. Use and properties of chayote (*Sechium edule*)

Sechium edule is a versatile cucurbitaceous crop valued for its nutritional, medicinal, agricultural, and industrial applications. Chayote has gained prominence for its edible parts, low caloric value, and wide adaptability to various agroclimatic conditions. Chayote is basically used for human consumption all over the world. Commonly, chayote is used as a vegetable and also used in stews or desserts. Fruit, stems, tender leaves, and tuberous parts of the adventitious roots are also eaten.

3.1 Nutritional and culinary uses

3.1.1 Fruit composition

Chayote fruit exhibits high moisture content (89.0–93.4%), contributing to its crisp texture and low energy density (26.0–31.0 kcal/100 g). It contains moderate carbohydrates (3.5–7.7%), primarily as soluble sugars (3.3%) and minimal starch (0.2%). Protein (0.9–1.1%) and fat (0.1–0.3%) contents are low, while dietary fiber ranges from 0.4–1.0%. Ash content (0.4–0.6%) reflects mineral presence, including calcium (12–19 mg), phosphorus (4–30 mg), and iron (0.2–0.6 mg). The fruit provides modest levels of B-vitamins—thiamine (0.03 mg), riboflavin (0.04 mg), niacin (0.4–0.5 mg)—and a moderate ascorbic acid content (11–20 mg). Vitamin A content is approximately 5 mg.

3.1.2 Seed composition

Often underutilized, the seed is nutritionally dense, with high carbohydrate content (~60%), including starch (1.9%) and soluble sugars (4.2%). It has a significantly higher protein content (5.5%) than the fruit, indicating potential for use as a plant-based protein source. Detailed data on fats, fiber, and micronutrients are limited, but available evidence supports their value as a functional food ingredient.

3.1.3 Young stem composition

Young stems contain higher protein (4.0%) and fiber (1.2%) than the fruit, along with moderate carbohydrates (4.7%), primarily as starch (0.7%) and minimal sugars (0.3%). Energy content is 60 kcal/100 g. Mineral concentrations are substantial: calcium (58 mg), phosphorus (108 mg), and iron (2.5 mg). Vitamin A is notably high (615 mg), along with thiamine (0.08 mg), riboflavin (0.18mg), niacin (1.1 mg), and ascorbic acid (16 mg), indicating strong nutritional and nutraceutical potential.

3.1.4 Root composition

Tuberous roots serve as the primary energy reserve, with the highest caloric content (79 kcal/100 g), driven by high starch (13.6%) and total carbohydrate (17.8%) levels. Protein (2.0%) and fat (0.2%) contents are moderate. Fiber (0.4%) and ash (1.0%) contents are low. Roots contain phosphorus (34 mg), calcium (7 mg), and iron (0.8 mg), and provide niacin (0.9 mg) and ascorbic acid (19 mg), supporting roles in energy metabolism and antioxidant function.

3.2 Medicinal properties

3.2.1 Pharmacological properties

The pharmacological activities of *Sechium edule* are primarily linked to its bioactive constituents, including both nutritional elements and nonnutritive nutraceuticals with strong antioxidant potential [43–45]. Flavonoids contribute to cardiovascular health by mitigating atherosclerosis and modulating lipid metabolism [46, 47]. Antihypertensive effects of hydroalcoholic extracts from roots, pulp, and peel are attributed to cinnamic acid methyl ester, coumaric acid, and vitexin [12].

Flavonoids and phenolic-rich polysaccharides also support glycemic control, as evidenced by improved glucose tolerance and reduced oxidative stress following chayote juice administration [48, 49]. Shoot extracts enhance AMPK activation and inhibit hepatic lipogenesis [50]. Polysaccharides, such as arabinans, homogalacturonans, arabinogalactans, and xylans, display immunomodulatory and antioxidant properties [51–53]. Crude pectins from fruit also exhibit cytoprotective effects [54].

Chayote demonstrates antiulcer activity by reducing gastric acidity and increasing pH [55]. Saponins from fruit and aerial parts exert antiproliferative effects against MK-1, HeLa, and B16F10 cells, while the seed-derived protein sechiumin possesses ribosome-inactivating and potential anticancer activity [56]. Extracts have been shown to induce apoptosis in acute myeloid leukemia cells [57] and exhibit antibacterial activity against *E. coli*, *K. pneumoniae*, *P. mirabilis*, and *E. cloacae* [58, 59]. However, the molecular mechanisms underlying these effects require further elucidation.

3.2.2 Antioxidant properties of Sechium edule

Sechium edule possesses notable antioxidant activity due to its rich phytochemical profile, including flavonoids, phenolic acids, and cucurbitacins. These compounds exert anti-inflammatory and antioxidative effects by enhancing the activity of endogenous enzymes such as superoxide dismutase (SOD) and glutathione peroxidase (GPx), which mitigate reactive oxygen species (ROS) accumulation.

Excess ROS can induce oxidative damage to DNA, lipids, and proteins, contributing to the development of chronic diseases such as metabolic syndrome. This condition is characterized by hyperglycemia, abdominal obesity, and hypertension, increasing the risk of cardiovascular disease and type 2 diabetes. Thus, regular consumption of antioxidant-rich foods like *S. edule* may support cellular defense mechanisms and reduce metabolic dysfunction.

3.2.3 Polyphenolic compounds

Polyphenols serve dual functions in plants, contributing to defense mechanisms against pathogens and herbivores, while offering various health benefits in humans, including antiallergic, anti-inflammatory, antiviral, anticancer, and antidiabetic effects [60–62]. Siciliano et al. [63] reported the highest total phenolic content in *Sechium edule* leaves (0.35 g/100 g DW), followed by roots (0.31 g/100 g DW) and stems (0.19 g/100 g DW). Notable compounds included flavonoids such as apigenin derivatives and diosmetin glycosides, with roots showing elevated levels of vicenin-2 (0.147 g/100 g DW) and vitexin (0.151 g/100 g DW), while leaves were rich in luteolin and apigenin glycosides.

Additional phenolics, including trans-cinnamic and phenylacetic acids, along with α-linolenic acid, were identified by Ragasa et al. [64]. Chao et al. [65] further showed that green chayote leaves had significantly higher total phenolic content (0.262 g GAE/100 g DW) than yellow varieties (0.063 g GAE/100 g DW).

3.2.4 Vitamins, carotenoids, and polysaccharides

Sechium edule is a rich source of essential vitamins and polysaccharides. The fruit is particularly high in folate (vitamin B9), important for neural tube defect prevention and pregnancy health, although its content significantly declines with cooking (from 93 to 18 μg/100 g DW). Vitamins A and E are most concentrated in the stems (615 IU and 90 mg/100 g DW, respectively), while the peel contains the highest levels of vitamin C (51.6 mg/100 g DW), followed by the fruit and leaves [66–68].

Carotenoids, including β-carotene and lutein, are present in various plant parts. The fruit peel contains 1.7 mg/100 g DW of total carotenoids and 0.36 mg/100 g DW of β-carotene [69]. Leaves are especially rich in lutein (7.4 mg/100 g FW) and β-carotene (4.4 mg/100 g FW) [70].

Chayote polysaccharides serve both structural and metabolic roles. Starch content ranges from 0.2–1.56 g/100 g DW in fruits and peaks in roots (13.6–72.8 g/100 g DW). Non-starch polysaccharides are abundant in peel (45.2 g/100 g DW), stems (21.7 g/100 g DW), and roots (16 g/100 g DW) [54]. Soluble fractions are mainly composed of arabinose, galactose, glucose, galacturonic acid, and mannose, while insoluble polysaccharides are dominated by galactose, with contributions from arabinose, xylose, and glucuronic acid.

3.2.5 Role in gastrointestinal and cardiovascular health

Sechium edule is a valuable source of dietary fiber, contributing to gastrointestinal and cardiovascular health. Its insoluble fiber increases fecal bulk and promotes bowel regularity, reducing the risk of constipation, hemorrhoids, and diverticulosis.

The soluble fiber fraction undergoes fermentation by gut microbiota, producing short-chain fatty acids (SCFAs) such as acetate, propionate, and butyrate. SCFAs support colonic epithelial health, reinforce mucosal integrity, and modulate inflammatory and immune responses. Systemically, SCFAs improve lipid metabolism and glycemic control, thereby lowering cardiovascular risk. Through these mechanisms, chayote enhances both gut microbiota function and host metabolic health.

4. Breeding and biotechnology of chayote

Sechium edule exhibits diverse fruit traits, including variations in shape, size, surface texture, color, and yield, highlighting its potential as a genetic resource [71]. However, limited availability of experimental materials and insufficient conservation efforts have impeded research and breeding progress. Challenges such as endocarpic and precocious seed germination, susceptibility to pests and diseases, and the need to preserve fruit quality complicate cultivar improvement. To address these issues, Newstrom [26] proposed the development of two breeding lines: one with unflavored fruits for industrial use and another with palatable fruits for culinary purposes. Aung et al. [28] recommended selection based on root traits (e.g., length and starch content) to identify disease-resistant genotypes suited to specific growth conditions.

4.1 Genetic improvement and hybridization in *Sechium edule*

Commercial chayote production is constrained by disease susceptibility and limited fruit quality, yet private breeding efforts remain minimal due to the plant's viviparous nature. Targeted breeding programs are needed to develop both unflavored cultivars for industrial processing and palatable ones for fresh consumption [26]. The cytogenetics of *S. edule* reveals a chromosome count of 2n = 28, with variations (e.g., 2n = 24) attributed to meroaneuploidy and chromosomal translocations.

Among the key cultivated botanical varieties, *virens levis* (smooth green chayote) dominates international markets, while others, such as *nigrum spinosum* and *nigrum xalapensis,* are regionally cultivated. Excessive clonal propagation in Costa Rica has resulted in genetic erosion, highlighting the need for renewed collection and evaluation of diverse landraces across Mexico and Central America [33].

A formal breeding initiative began in 1989 in Costa Rica's Valle de Ujarrás, focusing on traits such as fruit quality, pest and disease resistance, plant architecture, and yield components [72]. The program involved selecting 250 "mother plants," from which 34 were propagated asexually and characterized. Clonal lines such as EY13, SA15, and UJ28 demonstrated yields between 56.4 and 202.5 t/ha. Synthetic varieties—VS17, VS28, and VS29—exhibited good adaptability and fruit quality. Based on this success, germplasm from Costa Rica is now being introduced into other Central American countries, such as Honduras, Nicaragua, and Panama.

Inter- and intraspecific hybridization has been proposed but remains largely unexplored. Crosses between cultivated *S. edule* and wild relatives (*S. compositum*, *S. chinantlense*) show that gene flow among wild, semi-domesticated, and cultivated forms is ongoing, contributing to phenotypic diversity [72]. The cultivar 'Perla Negra', derived from *S. edule var. nigrum minor* and *S. edule var. amarus sylvestris*, demonstrated significant antiproliferative properties linked to bioactive compounds such as cucurbitacins and flavonoids [73].

Taxonomically, cultivated types are grouped under *S. edule ssp. edule*, while wild types fall under *S. edule* ssp. *sylvestris*, based on chromosomal and morphological criteria [74]. However, this classification lacks clarity due to potential intraspecific hybrids and spontaneous crosses, underscoring the need for deeper genomic and taxonomic resolution [75].

4.2 Molecular genetic markers

Molecular markers are essential tools for identifying genetic variation and associating traits with genomic loci [76–78]. Advances in marker technologies and sequencing have significantly enhanced crop improvement efforts [79–81]. These markers are broadly categorized into classical—morphological, cytological, biochemical—and DNA-based types, including RFLP, AFLP, SSRs, and SNPs [82, 83]. In *Sechium edule*, molecular markers have facilitated studies on genetic diversity and germplasm classification. For example, 42 Costa Rican accessions were characterized using isozyme markers [84], while 12 Indian accessions were grouped based on nine morphological traits and 20 RAPD markers [85]. Jain et al. [86] employed DAMD markers to assess variation across 36 Indian accessions, while Verma et al. [87] analyzed 74 landraces using RAPD and ISSR markers. These molecular tools provide a valuable foundation for trait mapping, genetic linkage analysis, and breeding strategies in *S. edule*.

4.3 Tissue culture

Plant tissue culture is a pivotal technique for clonal propagation and genetic improvement, enabling the regeneration of plants under sterile, controlled conditions [88]. In *Sechium edule*, successful regeneration has been achieved using explants such as nodal segments, leaves, petioles, shoots, and hypocotyls [89, 90]. Micropropagation protocols have employed Murashige and Skoog (MS) media supplemented with plant growth regulators, including 6-benzylaminopurine (BA), indole-3-acetic acid (IAA), gibberellic acid (GA_3), and indole-3-butyric acid (IBA), for callus induction, shoot regeneration, and rooting [91, 92].

Tissue culture has been instrumental in producing sterile seedlings to bypass seed propagation and reduce cultivation costs. It also provides a platform for genetic transformation *via Agrobacterium tumefaciens* or particle bombardment. Despite protocol development, transgenic *S. edule* lines remain largely unexplored.

5. Area and production of chayote

Sechium edule (chayote) is a Mesoamerican native crop historically cultivated by the Aztecs and Mayans [32]. The Spanish introduced it to Costa Rica, from where it spread throughout the Americas. Today, major producers include Mexico, Guatemala, Costa Rica (the top global exporter), the Dominican Republic, Brazil, and Peru. In the United States, chayote is grown in warmer states such as California, Florida, and Louisiana, although the majority of the supply is imported from Central America [93].

Globally, chayote is cultivated in parts of Asia and Europe, including India, Nepal, China, Taiwan, Papua New Guinea, Southeast Asia, Italy, and former Yugoslavia [28, 94, 95]. In India, it is mainly grown in Tamil Nadu (Nilgiris and Madurai), West Bengal, Himachal Pradesh, and the North Eastern Hill (NEH) region. Mizoram is the leading Indian producer, with 845 ha under cultivation, yielding 10,985 tons. Notably, high genetic diversity is observed in Mizoram, Meghalaya, and Sikkim, despite the crop's origin in Mexico [96].

Chayote is a high-volume but low-value crop across Central and South America and Southeast Asia. It is widely traded internationally, particularly to the U.S. and Europe [95]. In tropical Africa, it holds minor regional importance (e.g., Sierra Leone, Ghana, Malawi), though reliable production data are lacking. Brazil ranks chayote as its fifth most important commercial vegetable, with 170,000 tons produced in 1978; Mexico produced 12,000 tons in the same year. Costa Rica, despite lower output, is the leading exporter, shipping 4600 tons in 1982 valued at USD 1.5 million. In Indonesia's highlands, chayote is the most widely grown vegetable.

6. Variety development of chayote

The phenotypic diversity of *Sechium edule* (chayote) is considerable, with regional varieties differing in fruit morphology, epidermal characteristics, and organoleptic properties. The most widely cultivated form produces smooth, pale green, pear-shaped fruits. In Guatemala and Honduras, the 'Perulero' cultivar is characterized by a creamy outer skin, white flesh, and a softer, juicier consistency compared to the standard green type [97]. In the Central region of Veracruz (Mexico), local classifications group chayote into three types: white, green, and spiny, with the former two generally exhibiting smooth skin.

In the United States, predominant cultivars include 'Florida Green' and 'Monticello White,' typically characterized by smooth, non-spiny surfaces. Imported cultivars often exhibit rougher textures with grooves or trichomes [98]. Market diversity in North America reflects extensive hybridization and phenotypic selection, complicating taxonomic classification. A practical categorization by growers includes traits such as epidermal color (pale green, dark green, whitish), presence or absence of spines, surface texture (smooth vs. rough), and fruit shape (pear-shaped, globular, elongated) [93]. In India, no officially released cultivars exist, though imported strains such as 'Florida Green' and 'Monticello White' are occasionally grown. The ICAR Research Complex for the North Eastern Hill (NEH) Region, Umiam, Meghalaya, collected approximately 60 genotypes of *Sechium edule* from various states within the NEH region. These genotypes were evaluated for yield potential and quality traits. Certain accessions demonstrated high productivity, with some genotypes yielding up to 40 kg of fruit per plant [99]. Indian chayote exhibits a wide range of morphological traits, including variability in spine density (absent to dense), fruit shape (globular, elongated, flattened), and surface texture (smooth or rough) (**Table 2**) [97, 98].

In Ghana, a cultivar known as 'Ivory White' is cultivated, while the industrial sector typically prefers bland-flavored types, and consumers favor cultivars with enhanced taste and texture [99]. Wild chayote populations, endemic to moist

Strain	Fruit size	Fruit shape	Furrow depth	Spine density and location	Skin texture	Additional notes
BPI Ch1	Big	Elongated	Intermediate	Dense, entire fruit	Rough	—
BPI Ch2	Medium	Globular, slightly flat	Intermediate	Sparse, apex, and bottom	Rough	—
BPI Ch3	Medium	Globular, slightly flat	Shallow	Sparse, apex, and bottom	Smooth	—
BPI Ch4	Big	Elongated to long-elongated	Shallow	None to sparse, bottom	Smooth	—
BPI Ch5	Medium	Elongated	Shallow	Sparse, apex, and bottom	Rough	—
BPI Ch6	Medium	Globular, slightly flat	Shallow	Dense, entire fruit	Rough	—
BPI Ch7	Medium	Elongated	Shallow	None	Smooth	—
BPI Ch8	Medium	Globular to elongated	Intermediate	Sparse, apex, and bottom	Rough	—
BPI Ch9	Small	Elongated	Intermediate	Sparse, entire fruit	Rough	—
BPI Ch10	Small	Elongated	Intermediate	None to sparse, apex, and bottom	Smooth	—
BPI Ch11	Medium	Globular	Shallow	Dense and firm, entire fruit	Rough (dark green)	Recent introduction from Australia

Source: [97, 98].

Table 2.
Some chayote strains are.

environments of Mesoamerica (e.g., riverbanks, ravines), are morphologically similar to cultivated forms but display greater vegetative vigor, larger organs (leaves, flowers, and fruits), and more bitter, often spiny, dark green fruits [99].

7. Cultivation

7.1 Climate and soil

Sechium edule (chayote) is well adapted to tropical and subtropical climates and can be cultivated at elevations ranging from sea level up to 1800 meters above sea level (m asl) [100, 101]. For export-quality production, optimal growth has been reported at elevations between 1000 and 1200 m asl [102]. The crop performs best under daytime temperatures of 18–27°C (64 to 80°F) and nighttime temperatures above 15°C [101, 103]. Although chayote can tolerate a temperature range of 12–40°C, it is highly sensitive to frost and cannot survive, develop unripe fruit at temperatures below 12°C. Temperatures exceeding 28°C may lead to excessive vegetative growth and increased abscission of flowers and immature fruits, ultimately reducing yield. Chayote thrives in both full sun and partial shade, with a critical photoperiod of approximately 12 hours required to induce flowering [101, 103]. Optimal relative humidity for growth ranges between 80% and 85% [100]. Annual rainfall between 1500 mm and 2000 mm is considered ideal, although the plant can tolerate levels between 600 mm and 2500 mm [101]. However, chayote is intolerant of waterlogging, and proper drainage is essential. In regions with heavy rainfall, cultivation on raised beds or mounds is recommended to prevent root zone saturation [103].

Sechium edule (chayote) performs optimally in fertile, well-drained soils [101, 103]. Although the species can tolerate a soil pH range of 4.3 to 8.0, optimal growth occurs between pH 5.5 and 7.0 [101]. Cultivation in calcareous soils with high concentrations of calcium (Ca^{2+}) and magnesium (Mg^{2+}), such as those found in parts of Mexico, has been associated with reduced fruit yield and quality [104].

7.2 Propagation and planting

Chayote (*Sechium edule*) is primarily propagated via its large, recalcitrant seed enclosed within the fruit, which typically initiates germination upon reaching physiological maturity. To preserve viability, fruits can be stored at ~10°C for up to 6–8 weeks, delaying sprouting. Standard propagation involves direct planting of whole fruits, though commercial systems may employ rooted cuttings or selected seeds.

Cultivation is commonly performed on raised beds equipped with trellis or pandal systems to support vine growth, facilitate harvesting, and enhance postharvest handling. In sloped or hilly terrains, planting is synchronized with monsoon onset. Planting pits are amended with 10–15 kg of farmyard manure (FYM), and fruits are placed at a 45° angle with the stem-end slightly exposed to prevent rot. Deep planting increases susceptibility to decay. Commercial cultivation often includes chemical fertilizers, foliar nutrients, herbicides, and nematicides to improve yield and plant health.

Chayote (*Sechium edule*) is typically planted at a spacing of 3 × 2 meters in planting pits measuring 0.5 × 0.5 × 0.5 meters. Adequate irrigation is critical, especially during the dry season, due to the plant's shallow root system and high water requirements [105].

Sprinkler irrigation may offer dual effects: while it can reduce infestations of the red spider mite (*Tetranychus urticae* Koch), it may also interfere with pollinator activity and promote flower abscission [103, 105].

7.3 Manure and fertilizers

Chow-chow exhibits positive growth responses to proper manure and fertilizer application. Fertilizer requirements vary depending on soil type, climatic conditions, and cultivation practices. During land preparation, well-decomposed farmyard manure (FYM) should be incorporated at a rate of 15–20 tons per hectare. A recommended dose of 120:80:80 kg/ha of N: P_2O_5: K_2O is applied, with the full amounts of phosphorus and potassium incorporated prior to planting. Nitrogen is applied in split doses: half at the onset of vining and the remaining half prior to flowering.

7.4 Harvesting and yield

- *Harvesting-* Chayote (*Sechium edule*) fruits are typically selected based on morphological characteristics such as shape, color, external appearance, and the absence of physical damage or pest infestation [104]. Fruits are physiologically mature and ready for harvest approximately 35 days after pollination, at which point they typically weigh between 0.5 and 1.0 pounds and measure from 4 to 6 inches in diameter [106]. Harvesting is generally conducted in the early morning hours to take advantage of higher fruit turgidity and cooler ambient temperatures, which help maintain postharvest quality [104].

Harvesting is performed manually and is scheduled based on plant productivity. In commercial production systems, where plant material is more uniform, harvests are systematically planned, typically occurring from 2 to 4 times per week during the fruiting period. Harvests are usually divided, with one day dedicated to selecting export-quality fruit and the remaining days allocated for the domestic market.

- *Yield-* Chayote (*Sechium edule*) is a semi-perennial crop, with productive life cycles lasting from 2 to 3 years following establishment. Initial significant yields are typically observed beginning in the second year after planting. A single, well-maintained plant is capable of producing over 300 fruits annually, with individual fruit weights ranging from 200 to 450 grams.

Average fruit yields in commercial plantations range from 22 to 35 tons per hectare per year, though under optimal agronomic management, yields may reach approximately 40–50 tons per hectare (400–500 quintals/ha) [95]. In Northeast India, particularly in Meghalaya, slightly larger fruits are commonly observed, and yields of 80–100 quintals per acre have been recorded under favorable conditions.

7.5 Postharvest management

Postharvest handling of *Sechium edule* involves field-level sorting and packing into wooden crates for transport to packing facilities, where fruits undergo secondary grading. Based on market-specific standards (e.g., export vs. local), fruits are packed into cardboard boxes or plastic bags treated with antitranspirants and fungicides, then shipped in refrigerated containers to preserve quality [102, 107].

To prevent mechanical injuries, such as cuts and bruises that promote decay, careful handling is essential. Fruits are often individually wrapped in perforated polyethylene or waxed paper and stored at 7–13°C with 80–90% relative humidity [103, 104]. Storage below 7°C may cause chilling injury, while temperatures above 25°C can induce premature sprouting.

For commercial storage, pre-cooling via cold rooms or forced-air systems is standard. Packaging typically involves fiberboard boxes with internal dividers and individual film-wrapped fruits in a single layer to prevent compression damage. For household storage, fruits should be kept in paper bags within the refrigerator's vegetable compartment. Under optimal conditions, shelf life extends to 2–3 weeks, though mature fruits may sprout earlier and should be consumed sooner.

7.6 Plant protection

7.6.1 Pests and their management

Among the most damaging pests of *Sechium edule* is the red spider mite (Tetranychus urticae Koch), particularly prevalent during dry seasons. Infestation leads to foliar chlorosis and scab formation on fruits. Other acarid mites, including *Paraponychus corderoi* and *Eutetranychus banksi*, also cause significant damage [104].

Lepidopteran pests such as *Diaphania hyalinata* and *D. nitidalis* perforate immature fruits, increasing vulnerability to secondary fungal and bacterial infections. Sap-sucking Hemiptera—including *Aphis* spp. and *Bemisia tabaci*—impair photosynthetic efficiency and reduce fruit quality through surface blemishes. Coleopterans, such as *Phyllophaga* spp. and *Diabrotica* spp., damage both seedlings and fruits, adversely affecting plant vigor and yield [103, 107].

7.6.2 Diseases and their management

Plant-parasitic nematodes, notably *Heterodera radicicola* and *Meloidogyne incognita*, pose serious threats to root health by disrupting nutrient uptake, resulting in stunted growth and reduced vigor [103]. Fungal pathogens are the primary cause of chayote diseases. *Ascochyta phaseolorum* and *Fusarium oxysporum* can reduce commercial yields by 35–40% [104, 105], particularly during the rainy season when disease pressure is highest. Infections may occur at any plant growth stage. Excessive nitrogen fertilization has been associated with heightened susceptibility to *Mycovellosiela cucurbiticola* and *A. phaseolorum* [103].

Cultural control measures, such as manual removal of infected plant tissue, while effective in limiting pathogen spread, are labor-intensive. Crop rotation is recommended as an additional integrated pest management strategy to minimize pest and disease pressure.

8. Limitations, research gaps, and future prospects of *Sechium edule*

8.1 Biotic constraints

- *Pest and disease incidence*: *S. edule* faces persistent pest and pathogen pressures across growth stages, including postharvest.

- *Lack of resistant cultivars*: Breeding programs have not sufficiently exploited resistance traits from wild relatives or traditional landraces.
- *Absence of biocontrol strategies*: Integrated pest management lacks the use of biological control agents.

8.2 Genetic resource gaps

- *Underutilized genetic diversity*: Wild species and landraces remain largely untapped for breeding and trait improvement.
- *Limited germplasm access*: Conservation efforts are hampered by insufficient germplasm collections and poor maintenance.
- *Taxonomic ambiguity*: Phylogenetic relationships within *Sechium* require clarification through systematic taxonomy.

8.3 Socioeconomic and market constraints

- *Low market penetration*: Outside traditional regions, *S. edule* remains poorly recognized despite its multifaceted uses.
- *Commercial uniformity bias*: Market preferences for standardized fruit limit the use of morphologically diverse landraces.
- *Economic inequality*: Disparities between commercial and traditional growers undermine the in situ conservation of genetic diversity.

8.4 Conservation challenges

- *Dependence on traditional systems*: Conservation is largely sustained by traditional agriculture, which is threatened by modernization.
- *Habitat degradation*: Wild populations face an increasing risk due to deforestation and habitat loss, necessitating urgent conservation action.

8.5 Current initiatives

- *Conservation projects*: Ongoing efforts in Costa Rica and Nepal focus on both wild and cultivated forms, engaging multiple stakeholders.
- *Regional cooperation needed*: Cross-border collaboration, especially in Mesoamerica, is essential for germplasm exchange and coordinated conservation.

8.6 Research and utilization potential

- *Value-added applications*: Industrial use of fruit pulp and extraction of bioactive compounds (e.g., saponins) can enhance crop utility.
- *Medicinal and nutritional value*: Further research is needed on the phytochemical and nutraceutical properties.

- *Market diversification*: Promoting local varieties and by-products may create novel economic opportunities.

8.7 Strategic recommendations

- *Expand genetic research*: Systematic studies on wild and cultivated relationships are needed to support breeding and conservation.
- *Strengthen conservation efforts*: Prioritize in situ and ex situ strategies, particularly in biodiversity-rich regions like southern Mexico and Guatemala.
- *Policy and funding support*: Long-term success depends on political commitment and international collaboration, following models used for crops like cassava and rice.

Acknowledgements

The authors gratefully acknowledge the support and contributions of each co-author—Kedar Mahadev Gheware, Mukesh Kumar, Mohd. Wamiq, Manish Kumar, and Kushal Chaturvedi—in the successful completion of this book chapter.

Conflict of interest

The authors declare no conflict of interest.

Author details

Krishna Kaushik[1*], Kedar Mahadev Gheware[2], Mukesh Kumar[3], Mohd. Wamiq[4], Manish Kumar[1], Kushal Chaturvedi[1], Devanshu Shukla[3] and Saurabh Tiwari[5]

1 Department of Agriculture, Meerut Institute of Technology, Meerut, UP, India

2 PC Cell, AICRP on AZF, ICAR-Central Institute for Arid Horticulture, Bikaner, RJ, India

3 Department of Floriculture and Landscaping, SVPUA&T, Meerut, UP, India

4 Department of Agriculture, Sardar Patel University, Balaghat, MP, India

5 Department of Vegetable Science, SVPUA&T, Meerut, UP, India

*Address all correspondence to: kaushikkaali007@gmail.com

References

[1] Riviello-Flores MDLL, Arévalo-Galarza M, Lourdes D, Cadena-Iñiguez J, Soto-Hernández RM, et al. Nutraceutic characteristics of the extracts and juice of chayote (*Sechium edule* (Jacq.) Sw.) fruits. Beverages. 2018;**4**:e37. DOI: 10.3390/beverages4020037

[2] Taynath SJ, Singh BK, Jena S, Said PP. Optimization of process variables for petha manufacture from chayote (*Sechium edule* Sw.). Journal of Food Processing and Preservation. 2020;**44**:e14872. DOI: 10.1111/jfpp.14872

[3] Sudargo T, Aulia B, Prameswari AA, Isnansetyo A, Puspita ID, Budiyanti SA. Effect of administration of chaguro made of chayote (*Sechium edule*) and tuna (Thunnus sp.) on rats induced with streptozotocin nicotinamide and a high-fat diet. Current Research in Nutrition and Food Science Journal. 2021;**9**:258-266. DOI: 10.12944/CRNFSJ.9.1.24

[4] Vieira EF, Pinho O, Ferreira IMP, Delerue-Matos C. Chayote (*Sechium edule*): A review of nutritional composition, bioactivities and potential applications. Food Chemistry. 2018;275:557-568. DOI: 10.1016/j.foodchem.2018.09.146

[5] Bhat R, Ranjan K, Shabaraya AR. A comprehensive review on phytochemistry and pharmacological properties of *Sechium edule* (chayote). International Journal of Pharmacy and Pharmaceutical Sciences. 2025;**1**:1896-1902. DOI: 10.5281/zenodo.14718934

[6] Savinaya MS, Sangamesh SP, Narayana J, Krishna V. Traditional medicine knowledge and diversity of medicinal plants in Sharavathi valley region of Central Western Ghats. International Journal of Herbal Medicine. 2016;**4**(6):124-130

[7] Martins D, Nunez CV. Secondary metabolites from Rubiaceae species. Molecules. 2015;**20**(7):13422-13495

[8] Islam S, Kumar A, Dash KK, Alom S. Physicochemical analysis and nutritional properties of fresh, osmo-dehydrated and dried chayote (*Sechium edule* L.). Journal of Postharvest Technology. 2018;**6**(2):49-56. Available from: http://jpht.in/MenuscriptFile/da174c65-7dd9-4656-ae21-4674e86afd36.pdf [Accessed: April 21, 2021]

[9] Qiu Y, Liu G. Chayote—A potential vegetable crop for Florida: HS1454, 12/2022. EDIS. 2022;**2022**(6):1-10. DOI: 10.32473/edis-HS1454-2022

[10] Delgado-Durán R, Antonio-Pérez A, Esperón-Carreón J, Méndez-Guerrero OE, Saucedo SV, Rivero-Aranda RE, et al. Formulating chayote-enriched gummy candies: Balancing nutrition and taste for a delightful alternative. Food Chemistry Advances. 2025;**6**:100902. DOI: 10.1016/j.focha.2025.100902

[11] Lim T. *Sechium edule*. In: Lim T, editor. Edible Medicinal and Non-Medicinal Plants. Dordrecht: Springer; 2012. pp. 384-391. DOI: 10.1007/978-90-481-8661-7

[12] Vieira EF, Pinho O, Ferreira IMPLVO, Delerue-Matos C. Chayote (*Sechium edule*): A review of nutritional composition, bioactivities and potential applications. Food Chemistry. 2019;**275**:557-568

[13] Ke J, Jiang G, Shen G, Wu H, Liu Y, Zhang Z. Optimization, characterization

and rheological behavior study of pectin extracted from chayote (*Sechium edule*) using ultrasound-assisted method. International Journal of Biological Macromolecules. 2020;**147**:688-698. DOI: 10.1016/j.ijbiomac.2020.01.055

[14] Jena AK, Deuri R, Sharma P, Singh SP. Underutilized vegetable crops and their importance. Journal of Pharmacognosy and Phytochemistry. 2018;7:402-407

[15] Rosado-Pérez J, Aguiñiga-Sánchez I, Santiago-Osorio E, Mendoza-Núñez VM. Effect of *Sechium edule* var. nigrum spinosum (chayote) on oxidative stress and pro-inflammatory markers in older adults with metabolic syndrome: An exploratory study. Antioxidants. 2019;**8**:e146. DOI: 10.3390/antiox8050146

[16] Saenz F, Ruben R. Export contracts for non-traditional products: Chayote from Costa Rica. Journal on Chain and Network Science. 2004;**4**(2):139-150

[17] Pu YT, Luo Q, Wen LH, Li YR, Meng PH, Wang XJ, et al. Origin, evolution, breeding, and omics of chayote, an important cucurbitaceae vegetable crop. Frontiers in Plant Science. 2021;**12**:739091

[18] Rosas EP, Reyes A, Castro-Olivera PM, Acosta-Nava K, Vega-Goitia M, Palacios A. Food drying: Opportunities, market solutions, and nutraceutical properties of chayote. Alim Cienc Ing. 2023;**30**(2):1-21

[19] Machida-Hirano R, Cortés-Cruz M, González BAA, Íñiguez JC, Shirata K, Watanabe KN. Isolation and characterization of novel microsatellite markers in chayote [*Sechium edule* (Jacq.) Sw.]. American Journal of Plant Sciences. 2015;**6**(13):2033

[20] Tchamgoue J, Tchokokam YRW, Ngouonpe AW, Tchangoue YAN, Tiani LM, Msagati TAM, et al. The genus canthium: A comprehensive summary on its traditional use, phytochemistry, and pharmacological activities. Fitoterapia. 2023;**172**(3):105754

[21] Goldblatt P. Index to plant chromosome numbers (1975-1978). In: Monographs in Systematic Botany from the Missouri Botanical Garden. Vol. 33. No. 4. Missouri: Brittonia; 1981. p. 589

[22] Goldblatt P, editor. Index to plant chromosome numbers (1979-1981). In: Monographs in Systematic Botany from the Missouri Botanical Garden. Vol. 8. Missouri: Monographs Systematic Botany; 1984. p. 427

[23] Goldblatt P. Index to Plant chromosome numbers 1979-1981. Monographs in Systematic Botany from the Missouri Botanical Garden. 1983;**8**:212-222

[24] Giusti L, Resnik M, Ruiz TV, Grau A. Notes on the biology of *Sechium edule* (Jacq.) Swartz (cucurbitaceae). Lilloa. 1978;**35**:5-13

[25] Singh AK. Cytogenetics and evolution in the cucurbitaceae. In: Bates DM, Robinson RW, Jeffrey C, editors. Biology and Utilization of the Cucurbitaceae. Ithaca, NY: Cornell University Press; 1990. pp. 10-28

[26] Newstrom LE. Studies in the origin and evolution of chayote, *Sechium edule* (Jacq.) Sw. (cucurbitaceae) [PhD thesis]. Berkeley: University of California; 1986. 149 pp

[27] Ramírez MA, Valverde E, Sáenz MV. Study of some factors that affect the fertilization of flowers and the abstention of chayote fruits (*Sechium edule* Sw.) in cultivation in Costa Rica. Turrialba. 1990;**40**:340-345

[28] Aung LH, Ball A, Kushad M. Developmental and nutritional aspects of chayote (*Sechium edule*, Cucurbitaceae). Economic Botany. 1990;**44**:157-164

[29] Cook OF. The chayote: A tropical vegetable. USDA Division Botany Bulletin. 1901;**18**:1-31

[30] Newstrom LE. Evidence for the origin of chayote *Sechium edule* (Cucurbitaceae). Economic Botany. 1991;**45**:410-428

[31] León J. Fundamentos Botánicos de los Cultivos Tropicales. OEA, San José, Costa Rica: Instituto Interamericano de Ciencias Agrícolas; 1968

[32] Bukasov SM. Las Plantas Cultivadas de México, Guatemala y Colombia. In: CATIE/GTZ, Turrialba, Costa Rica. 1st ed. 1981. [In Spanish] (J. León, Transl.). 173 pp

[33] Engels JMM. Variation in *Sechium edule* Sw. in Central America. Journal of the American Society for Horticultural Science. 1983;**108**:706-710

[34] Maffioli A. Recursos Genéticos de Chayote, Sechium edule (Jacq.) Swartz. (Cucurbitaceae). CATIE/GTZ, Turrialba: Costa Rica; 1983. p. 151

[35] Cruz-León A, Querol D. Catálogo de Recursos Genéticos de Chayote (*Sechium edule* Sw.) en el Centro Regional Universitario Oriente de la Universidad Autónoma de Chapingo. Chapingo: Universidad Autónoma de Chapingo; 1985. 235 pp

[36] Lira R. Estudios Taxonómicos y Ecogeográficos de las Cucurbitaceae Latinoamericanas de Importancia Económica: Cucurbita, Sechium, Sicana y Cyclanthera. In: Systematic and Ecogeographic Studies on Crop Genepools, No. 9. Rome, Italy: International Plant Genetic Resources Institute; 1995. 281 pp

[37] Cruz-León A. ¿Chayote o cruzas intergenéricas? Hallazgo y características. Rev Geogr Agr. 1986;**9-10**:100-106

[38] Lira R, Chiang F. Two new combinations in Sechium (Cucurbitaceae) from Central America and a new species from Oaxaca, Mexico. Novon. 1992;**2**:227-231

[39] Backer CA, Bakhuizen van der Brink RC. Flora of Java. Vol. 1. Noordhoff, Groningen, The Netherlands: N.V.P; 1963

[40] Cordenoy EJ. Flore de l'île de la Réunion. Paris: P. Klinsksieck; 1895

[41] Brücher H. Useful Plants of Neotropical Origin and their Wild Relatives. New York: Springer Verlag; 1989

[42] Jeffrey C. Appendix: An outline classification of the Cucurbitaceae. In: Bates DM, Robinson RW, Jeffrey C, editors. Biology and Utilization of the Cucurbitaceae. Ithaca, NY: Cornell University Press; 1990. pp. 449-463

[43] Meisel H. Biochemical properties of bioactive peptides derived from milk proteins: Potential nutraceuticals for food and pharmaceutical applications. Livestock Production Science. 1997;**50**:125-138. DOI: 10.1016/S0301-6226(97)00083-3

[44] Nasri H, Baradaran A, Shirzad H, Rafieian-Kopaei M. New concepts in nutraceuticals as alternative for pharmaceuticals. International Journal of Preventive Medicine. 2014;**5**:1487-1499

[45] Santini A, Novellino E. To nutraceuticals and back: Rethinking a

concept. Food. 2017;**6**:e74. DOI: 10.3390/foods6090074

[46] Juan D, Pérez-Vizcaíno F, Jiménez J, Tamargo J, Zarzuelo A. Flavonoids and cardiovascular diseases. Studies in Natural Products Chemistry. 2001;**25**:565-605. DOI: 10.1016/S1572-5995(01)80018-1

[47] García-Lafuente A, Guillamón E, Villares A, Rostagno MA, Martínez J. Flavonoids as anti-inflammatory agents: Implications in cancer and cardiovascular disease. Inflammation Research. 2009;**58**:537-552. DOI: 10.1007/s00011-009-0037-3

[48] Mukherjee PK, Nema NK, Maity N, Sarkar BK. Phytochemical and therapeutic potential of cucumber. Fitoterapia. 2013;**84**:227-236. DOI: 10.1016/j.fitote.2012.10.003

[49] Loizzo MR, Bonesi M, Menichini F, Tenuta MC, Leporini M, Tundis R. Antioxidant and carbohydrate-hydrolysing enzymes potential of *Sechium edule* (Jacq.) Swartz (Cucurbitaceae) peel, leaves and pulp fresh and processed. Plant Food for Human Nutrition. 2016;**71**:381-387. DOI: 10.1007/s11130-016-0571-4

[50] Wu CH, Ou TT, Chang CH, Chang XZ, Yang MY, Wang CJ. The polyphenol extract from *Sechium edule* shoots inhibits lipogenesis and stimulates lipolysis via activation of AMPK signals in HepG2 cells. Journal of Agricultural and Food Chemistry. 2014;**62**:750-759. DOI: 10.1021/jf404611a

[51] Ordonez A, Gomez J, Vattuone M. Antioxidant activities of *Sechium edule* (Jacq.) Swartz extracts. Food Chemistry. 2006;**97**:452-458. DOI: 10.1016/j.foodchem.2005.05.024

[52] Duan J, Dong Q, Kan D, Fa NJ. Characterization of a pectic polysaccharide from the leaves of diospyros kaki and its modulating activity on lymphocyte proliferation. Biopolymers. 2010;**93**:649-656. DOI: 10.1002/bip.21430

[53] Castro-Alves VC, Do Nascimento JRO. Polysaccharides from raw and cooked chayote modulate macrophage function. Food Research International. 2016;**81**:171-179. DOI: 10.1016/j.foodres.2016.01.017

[54] Shiga TM, Peroni-Okita FHG, Carpita NC, Lajolo FM, Cordenunsi BR. Polysaccharide composition of raw and cooked chayote (*Sechium edule* Sw.) fruits and tuberous roots. Carbohydrate Polymers. 2015;**130**:155-165. DOI: 10.1016/j.carbpol.2015.04.055

[55] Sateesh G, Hussaini SF, Kumar GS, Rao B. Anti-ulcer activity of *Sechium edule* ethanolic fruit extract. Pharma Innovation. 2012;**5**:77-81

[56] Tamayo JF, Peña GR, De La Paz LC, Pacheco CH, Campusano YA. *Sechium edule* (Jacq.) Sw: Potencia fitoterapéutica Como agente antibacteriano. MediSur. 2016;**14**:664-670

[57] Aguiñiga-Sánchez I, Soto-Hernández M, Cadena-Iñiguez J, Ruiz-Posadas LDM, Cadena-Zamudio JD, González-Ugarte AK, et al. Fruit extract from a *Sechium edule* hybrid induce apoptosis in leukaemic cell lines but not in normal cells. Nutrition and Cancer. 2015;**67**:250-257. DOI: 10.1080/01635581.2015.989370

[58] Ordoñez AAL, Gómez JD, Cudmani NM, Vattuone MA, Isla MI. Antimicrobial activity of nine extracts of *Sechium edule* (Jacq.) Swartz. Microbial Ecology in Health and Disease. 2003;**15**:33-39. DOI: 10.1080/0891060010015583

[59] Zampini IC, Cuello S, Alberto MR, Ordoñez RM, Almeida RD, Solorzano E. Antimicrobial activity of selected plant species from "the Argentine Puna" against sensitive and multi-resistant bacteria. Journal of Ethnopharmacology. 2009;**124**:490-505. DOI: 10.1016/j.jep.2009.05.011

[60] Kratchanova M, De Nev P, Ciz M, Lojek A, Mihailov A. Evaluation of antioxidant activity of medicinal plants containing polyphenol compounds. Acta Biochimica Polonica. 2010;**57**:229-234. DOI: 10.1002/rhc3.24

[61] Chun-Lin KE, Wang D, Xie HW, Wang ZG, Chao-Tian LV, Deng YX. Research progress in antibacterial activity of polyphenol compounds. Science and Technology of Food Industry. 2012;**17**:364-367. DOI: 10.13386/j.issn1002-0306.2012.17.057

[62] Lattanzio V. Phenolic Compounds: Introduction. Berlin: Springer; 2013

[63] Siciliano T, Tommasi N, Morelli I, Braca A. Study of flavonoids of *Sechium edule* (Jacq.) Swartz (Cucurbitaceae) different edible organs by liquid chromatography photodiode array mass spectrometry. Journal of Agricultural and Food Chemistry. 2004;**52**:6510-6515. DOI: 10.1021/jf040214q

[64] Ragasa CY, Biona K, Shen CC. Chemical constituents of *Sechium edule* (Jacq.) Swartz. Der Pharma Chemica. 2014;**6**:251-255

[65] Chao PY, Lin SY, Lin KH, Liu YF, Ju-Ing H, Yang CM, et al. Antioxidant activity in extracts of 27 indigenous Taiwanese vegetables. Nutrients. 2014;**6**:2115-2130. DOI: 10.3390/nu6052115

[66] Kumar A, Islam S, Dash KK, Sarkar S. Optimization of process parameters for osmotic dehydration of chayote cubes by response surface methodology. International Journal of Agriculture Environment and Biotechnology. 2017;**10**:725-737. DOI: 10.5958/2230-732X.2017.00090.0

[67] Fadliya F, Supriadi S, Diah AWM. Analisis vitamin C dan protein pada biji buah labu siam (*Sechium edule*). Jurnal Akademika Kimia. 2018;**7**:6-10. DOI: 10.22487/j24775185.2018.v7.i1.10383

[68] Daulay A, Syahputra R, Nafitri A. Antioxidant activity test of chayote (*Sechium edule* (Jacq.) Swartz) ethanol extract using DPPH method. Journal de Physique. 2021;**1819**:012035. DOI: 10.1088/1742-6596/1819/1/012035

[69] Medina TDS. Extraction and characterization by UPLC–MSE of phenolic compounds obtained from vegetable waste in food services [thesis]. Federal University of the State of Rio De Janeiro-Unirio; 2019

[70] Sriwichai W, Berger J, Picq C, Avallone S. Determining factors of lipophilic micronutrient bioaccessibility in several leafy vegetables. Journal of Agricultural and Food Chemistry. 2016;**64**:1695-1701. DOI: 10.1021/acs.jafc.5b05364

[71] Newstrom LE. Origin and evolution of chayote, *Sechium edule*. In: Bates DM, Robinson RW, Jeffrey C, editors. Biol Utiliz Cucurbitaceae. 1st ed. 2019. pp. 141-149. DOI: 10.7591/9781501745447-014

[72] Brenes-Hine A. Proyecto: Conservación de germoplasma de chayote [*Sechium edule* (Jacq.) Swartz] y tacaco [Sechium tacaco (Pittier) C. Jeffrey] como una base de apoyo para el mejoramiento genético y la producción de semillas [thesis]. Heredia: Universidad Nacional; 2002

[73] Salazar-Aguilar S, Ruiz-Posadas LM, Cadena-Iñiguez J, Soto Hernández M, Santiago-Osorio E, Aguiñiga-Sánchez I, et al. *Sechium edule* (Jacq.) Swartz, a new cultivar with antiproliferative potential in a human cervical cancer HeLa cell line. Nutrients. 2017;**9**(8):798

[74] Lira R, Castrejon J, Zamudio S, Rojas-Zenteno C. Proposed taxonomic location for the wild chayotes (*Sechium edule*, Cucurbitaceae) of Mexico. Acta Botanica Mexicana. 1999;**49**:47-61. DOI: 10.21829/abm49.1999.838

[75] Gamboa W. Producción agroecológica: una opción para el desarrollo del cultivo del chayote (*Sechium edule* (Jacq.) Sw.) [thesis]. Costa Rica: Universidad de Costa Rica; 2005

[76] Domoney C, Duc G, Ellis TN, Ferrándiz C, Firnhaber C, Gallardo K, et al. Genetic and genomic analysis of legume flowers and seeds. Current Opinion in Plant Biology. 2006;**9**:133-141. DOI: 10.1016/j.pbi.2006.01.014

[77] Duran C, Appleby N, Edwards D, Batley J. Molecular genetic markers: Discovery, applications, data storage and visualisation. Current Bioinformatics. 2009;**4**:16-27. DOI: 10.2174/157489309787158198

[78] Pérez-de-Castro AM, Vilanova S, Cañizares J, Pascual L, Blanca JM, Díez MJ, et al. Application of genomic tools in plant breeding. Current Genomics. 2012;**13**:179-195. DOI: 10.2174/138920212800543084

[79] Wani SN, Ahmad MF, Malik RA. Genetic markers: A tool for genetic improvement in fruit crops. Progressive Horticulture. 2007;**39**:5-16

[80] Bhat ZA, Dhillon WS, Rashid R, Bhat JA, Dar WA, Ganaie MY. The role of molecular markers in improvement of fruit crops. Notulae Scientia Biologicae. 2010;**2**:22-23. DOI: 10.15835/nsb224222

[81] Hawkesford MJ. The diversity of nitrogen use efficiency for wheat varieties and the potential for crop improvement. Better Crop Plant Food. 2012;**96**:10-12

[82] Lewis RS, Nicholson JS, Moon HS. Use of transferable *Nicotiana tabacum* L. microsatellite markers for investigating genetic diversity in the genus Nicotiana. Genome. 2008;**51**:547-559. DOI: 10.1139/G08-039

[83] Wang YJ, Wang CY, Hong Z, Hao L, Liu XL, Ji WQ. Genetic resources and crop evolution. Genetic Resources and Crop Evolution. 2013;**60**:115-117. DOI: 10.1007/s10722-012-9907-6

[84] Vitomir V, Milanko S, Dragomir L, Ljuba Š. Some aspects of implementation DNA markers in animal breeding. Letopis Naunih Radova. 2012;**36**:101-108

[85] Joshi BK, Shrestha S, Adhikri B, Bhattarai M. Traditional practices and genetic diversity on chayote landraces and their conservation. Sustainable Development. 2020;**10**:272-288. DOI: 10.31924/nrsd.v10i2.060

[86] Jain JR, Timsina B, Satyan KB, Manohar SH. A comparative assessment of morphological and molecular diversity among *Sechium edule* (Jacq.) Sw. Accessions in India. 3 Biotech. 2017;**7**:e106. DOI: 10.1007/s13205-017-0726-5

[87] Verma VK, Pandey A, Jha AK, Ngachan SV. Genetic characterization of chayote [*Sechium edule* (Jacq.) Swartz.] landraces of north Eastern Hills of India and conservation measure. Physiology and Molecular Biology of

Plants. 2017;**23**:911-924. DOI: 10.1007/s12298-017-0478-z

[88] Kyte L, Kleyn J, Scoggins H, Bridgen M. Plants from Test Tubes: An Introduction to Micropropagation. Portland: Timber Press; 2013. p. 250

[89] Abdelnour-Esquivel A, Engelmann F. Cryopreservation of chayote (*Sechium edule* Jacq. Sw.) zygotic embryos and shoot-tips from in vitro plantlets. CryoLetters. 2002;**23**:299-308. DOI: 10.1016/S1095-6433(02)00130-7

[90] Cruz-Martinez V, Castellanos-Hernandez OA, Acevedo-Hernandez GJ, Torres-Moran MI, Gutierrez-Lomeli M, Ruvalcaba-Ruiz D, et al. Genetic fidelity assessment in plants of *Sechium edule* regenerated via organogenesis. South African Journal of Botany. 2017;**112**:118-122. DOI: 10.1016/j.sajb.2017.05.0200

[91] Katase M. Factors affecting proliferation of shoot in mulberry axillary bud culture. Journal of Sericultural Science of Japan. 1993;**62**:152-161. DOI: 10.11416/kontyushigen1930.62.152

[92] Amilineni U, Gangal V, Gudasalamani R, Karaba NN, Ramanan US. Establishment and standardization of in vitro regeneration protocol in Nothapodytes nimmoniana Graham and evaluation of camptothecine (CPT) in tissue culture plants. Indian Journal of Plant Physiology. 2016;**21**:1-7. DOI: 10.1007/s40502-015-0182-3

[93] Anonymous. Characteristics of chayote [Internet]. 2021. Available from: https://www.botanical-online.com/en/botany/chayote-plant-characteristics

[94] Chakravarty HL. Cucurbits of India and their role in the development of vegetable crops. In: Bates DM, Robinson RW, Jeffrey C, editors. Biology and Utilization of the cucurbitaceae. Ithaca, NY: Cornell University Press; 1990. pp. 325-334

[95] Engels JMM, Jeffrey C. *Sechium edule* (Jacq.) Swartz. In: Siemonsma JS, Piluek K, editors. Plant Resources of South-East Asia. Vol. 8. Vegetables. Wageningen: Pudoc Scientific Publishers; 1993. pp. 246-248

[96] Rai N, Yadav RK, Yadav DS. Genetic variability studies in chow-chow (*Sechium edule* SW). Vegetable Science. 2005;**32**:184-186

[97] Tibor Ra JG. Chayote growing [Internet]. 2017. Available from: https://www.facebook.com/notes/philippine-association-of-urban-agriculturists/chayote-growing/1926785537609164

[98] Magsasaka J. Chayote (Sayote) cultivation guide: All you need to know in planting, growing and harvesting chayote [Internet]. 2021. Available from: https://www.juanmagsasaka.com/2020/12/chayote-sayote-cultivation-guide-all.html

[99] Engels JMM. *Sechium edule* (Jacq.) Sw. PROTA4U [Internet]. 2021. Available from: https://www.prota4u.org/database/protav8.asp?fr=1&g=pe&p=Sechium+edule+(Jacq.)+Sw

[100] Rojas-Sandoval J. *Sechium edule* (Chayote). CABI compendium: Invasive species. Wallingford, UK: CAB International; 2020. DOI: 10.1079/ISC.49493.20203482792

[101] Fern K. Sechium edule (Jacq.) Sw. Tropical Plants Database [Internet]. 2020;**27**. Available from: http://tropical.theferns.info/viewtropical.php?id=Sechium+edule [Accessed: July 15, 2020]

[102] Flores ME. El Chayote, *Sechium edule* Swartz (Cucurbitaceae). Revista de Biología Tropical. 1989;**37**:1-54

[103] Valenzuela H. Chayote Production Guidelines for Hawaii. University of Hawaii, College of Tropical Agriculture and Human Resources Vegetable Crops Extension [Internet]. 2020. Available from: http://www2.hawaii.edu/~hector/prod%20guides%20fold/CHAYOTE.pdf [Accessed: July 15, 2020]

[104] Cadena-Iñiguez J, Soto-Hernández M, Torres-Salas A, Aguiñiga-Sánchez I, Ruiz-Posadas L, Rivera-Martínez AR, et al. The antiproliferative effect of chayote varieties (*Sechium edule* (Jacq.) Sw.) on tumor cell lines. Journal of Medicinal Plant Research: Planta Medica. 2013;7(8):455-460

[105] Lira Saade R. Chayote (*Sechium edule*). 2021. Available from: https://hort.purdue.edu/newcrop/1492/chayote.html

[106] University of California Master Gardeners, Santa Clara County, CA. Chayote. University of California, Division of Agriculture and Natural Sciences [Internet]. 2021. Available from: http://mgsantaclara.ucanr.edu/garden-help/vegetables/chayote [Accessed: January 20, 2021]

[107] Lira SR. Chayote, *Sechium edule* (Jacq.) Sw. Promoting the Conservation and Use of Underutilized and Neglected Crops. 8. Institute of Plant Genetics and Crop Plant Research/International Plant Genetic Resource Institute: Gatersleben/Rome; 1996. 70 p

Chapter 3

Harnessing Indigenous Cucurbits for Food and Nutrition Security in Kenya: Focus on Tropical Pumpkin, Horned Melon, and Bottle Gourd

Samuel Linguya Kimaru, William Maina Muiru and John Wangai Kimenju

Abstract

In Kenya, members of the Cucurbitaceae family such as tropical pumpkin (*Cucurbita moschata*), horned melon (*Cucumis metuliferus*), and bottle gourd (*Lagenaria siceraria*) play important roles in rural agriculture, nutrition, and traditional livelihoods. This study evaluated 15 accessions of cucurbit crops—tropical pumpkin, horned melon, and bottle gourd—collected from diverse regions across Kenya. The primary objective was to identify high-performing accessions suitable for breeding, regional adaptation, and commercialization while also examining market dynamics. Field trials were conducted using a randomized complete block design, and key crop traits were analyzed. Varied statistical tools were applied to evaluate trait variability and relationships among accessions. The results revealed significant differences ($p \leq 0.05$) across accessions, with Tharaka-HML-03, Embu-PMK-01, and Meru-BLG-05 standing out in vine length and fruit production. Principle component analysis (PCA) identified fruit weight and vine length as major contributors to phenotypic variation, while clustering indicated distinct diversity patterns, especially within tropical pumpkin. Market surveys highlighted that tropical pumpkin was the most accessible and affordable crop, whereas horned melon, though scarce, fetched premium prices due to its novelty and health appeal. Bottle gourd was moderately available and most affordable in regions like Meru. Consumer preferences focused on taste, tenderness, and shelf life, while traders valued visual traits and uniformity. The study concludes that integrating agronomic data with market insights provides a solid foundation for breeding programs and commercialization strategies. Recommendations include multilocation trials, enhanced postharvest handling, consumer-focused breeding, and targeted farmer training to promote cucurbit production and value chain development across Kenya.

Keywords: accessions, commercialization, cucurbit, dendrogram, morphological

1. Introduction

1.1 Background information

Cucurbit crops are an integral part of global agriculture, particularly in tropical and subtropical regions where they are cultivated for their nutritional and economic importance. In Kenya, the major cucurbits include pumpkin (*Cucurbita maxima*), butternut (*Cucurbita moschata*), watermelon (*Citrullus lanatus*), cucumber (*Cucumis sativus*), and courgette (*Cucurbita pepo*), which are grown widely for domestic use and export [1, 2]. These crops are rich sources of vitamins, minerals, and dietary fiber [3]. Recent research has highlighted the potential of semi-bush (SB) pumpkin genotypes in intercropping systems, particularly with short-season legumes like beans (*Phaseolus vulgaris*) and cowpeas (*Vigna unguiculata*) [4]. This approach not only improves land-use efficiency but also offers higher overall productivity without compromising the yield of the main pumpkin crop [4, 5].

Horned melon (*Cucumis metuliferus*), commonly known as thorn melon, is a lesser-known cucurbit native to Africa but now gaining international attention due to its adaptability and nutritional profile [5, 6]. The fruit contains high levels of phytochemicals, vitamins, and antioxidants and is also valued for its high water content, making it ideal for arid and semiarid environments [7, 8]. Though edible raw, it can also be processed into juices, yoghurts, and even ice creams [9, 10]. Its decorative appeal further enhances its marketability. Despite these advantages, horned melon remains underutilized due to limited awareness, low consumer acceptance, and insufficient research on its agronomic traits and genetic diversity in Kenya [6, 11].

Cucumber (*Cucumis sativus*), one of the oldest cultivated crops, originated in ancient Egypt and is now a daily staple in many regions, particularly Africa and Asia [12]. It is a rich source of moisture, minerals, and bioactive phytochemicals that support food quality and human health [5, 13]. Along with related species such as horned melon and gherkin (*Cucumis anguria*), cucumber has shown promising adaptability beyond its traditional growing zones [13]. With increasing interest in functional foods and natural health products, cucurbits are now being explored for their potential as nutraceuticals and sources of bioactive compounds from both edible and non-edible parts [7, 13]. This trend supports a circular economy by allowing for the valorization of food waste and increasing the sustainability of food systems [10, 13]. Strategic research, value chain development, and crop improvement efforts are essential to unlock the full potential of these crops.

1.2 Production of tropical pumpkin, horned melon, and bottle gourd in Kenya

1.2.1 Plant taxonomy

1.2.1.1 Tropical pumpkin (Cucurbita moschata Duchesne)

Tropical pumpkin, botanically known as *Cucurbita moschata*, belongs to the family Cucurbitaceae, which consists of herbaceous vines native to tropical and subtropical regions. This species is one of the five domesticated *Cucurbita* species, which include *C. pepo*, *C. maxima*, *C. argyrosperma*, and *C. ficifolia* [2]. *C. moschata* is characterized by its trailing vines, large leaves with coarse hairs, and yellow, unisexual flowers. The fruit is variable in size, shape, and color, often possessing thick rinds and sweet, orange flesh rich in beta-carotene [1]. It is especially suited to hot, humid climates and shows better tolerance to high temperatures and disease pressure compared to *C. maxima* and *C. pepo* [1]. In Africa, it is a staple crop and is valued for both its nutritional content and long storage potential (**Figure 1**).

1.2.1.2 Horned melon (Cucumis metuliferus E. Mey. ex Naudin)

Horned melon is a member of the genus *Cucumis*, within the family Cucurbitaceae. It is indigenous to sub-Saharan Africa and has adapted well to various agroecological zones, including semiarid regions [11]. Horned melon is a monoecious or dioecious annual climber that produces distinctive orange-skinned fruits with horn-like spines and green, jelly-like pulp filled with edible seeds [14]. This species is noted for its high drought tolerance and its ability to grow in marginal soils with low fertility [6]. Beyond its traditional consumption in rural Africa, it has attracted global attention for its ornamental value and high levels of bioactive compounds such as phenolics, flavonoids, and antioxidants [7]. Its underutilization has spurred calls for conservation and genetic improvement to fully exploit its potential (**Figure 2**).

1.2.1.3 Bottle gourd (Lagenaria siceraria (Molina) Standl.)

Bottle gourd is an ancient domesticated crop that belongs to a different genus within the Cucurbitaceae family. Known for its diverse fruit morphology, ranging from bottle-shaped, round, to elongated shape, it is used for both food and utility purposes such as utensils and musical instruments when dried [3]. It grows as a fast-growing, trailing, or climbing vine with white, funnel-shaped flowers and large fruits

Figure 1.
*Tropical pumpkin (*Cucurbita moschata*) plant and inflorescence.*

Figure 2.
*Horned melon (*Cucumis metuliferus*) plant and fruit.*

that can weigh several kilograms. While the tender fruits are consumed as vegetables, mature fruits are dried and used as containers or crafts. The crop is highly adaptive, with tolerance to drought and poor soils, making it well-suited for cultivation in arid and semiarid regions [15]. Due to its cultural, nutritional and economic importance, it is considered a multipurpose species with potential for crop improvement (**Figure 3** and **Table 1**).

1.2.2 Production of tropical pumpkin, horned melon, and bottle gourd

The cultivation of tropical pumpkin (*Cucurbita moschata*), horned melon (*Cucumis metuliferus*), and bottle gourd (*Lagenaria siceraria*) is gaining increasing attention across Africa, especially in Kenya, due to their nutritional value and economic potential. These crops play an important role in supporting food security and rural incomes. Tropical pumpkin, in particular, holds a prominent place in Kenyan agriculture, with an estimated annual production of 300,000 tons, meeting both household consumption and market demands (Food and Agriculture Organization) [16, 17]. Its widespread availability has contributed significantly to nutrition and market systems within the country.

Horned melon remains a relatively new and underutilized crop in Kenya, though interest in its cultivation is growing, particularly in the Rift Valley region. While production data is still limited, the fruit's increasing popularity, driven by awareness of

Figure 3.
*Bottle gourd (*Lagenaria siceraria*) inflorescence and fruits.*

Taxonomic rank	**Tropical pumpkin (*Cucurbita moschata*)**	**Horned melon (*Cucumis metuliferus*)**	**Bottle gourd (*Lagenaria siceraria*)**
Kingdom	Plantae	Plantae	Plantae
Phylum (division)	Magnoliophyta (Angiosperms)	Magnoliophyta (Angiosperms)	Magnoliophyta (Angiosperms)
Class	Magnoliopsida (Eudicots)	Magnoliopsida (Eudicots)	Magnoliopsida (Eudicots)
Order	Cucurbitales	Cucurbitales	Cucurbitales
Family	Cucurbitaceae	Cucurbitaceae	Cucurbitaceae
Genus	*Cucurbita*	*Cucumis*	*Lagenaria*
Species	*Cucurbita moschata*	*Cucumis metuliferus*	*Lagenaria siceraria*

Table 1.
Taxonomy of selected cucurbits.

its health benefits, presents a promising opportunity for agribusiness development. In contrast, bottle gourd is a well-established crop in Kenya, appreciated for its culinary uses and medicinal value. Its estimated production stands at 250,000 tons annually [16, 17], positioning Kenya as a notable producer in Africa, though behind global leaders like India. The continued cultivation and market development of these three crops are vital for strengthening Kenya's agricultural sector. Investments in production practices, market infrastructure, and public education could further enhance the sustainability and economic impact of these crops.

1.2.3 Climate and ecological requirements

The ecological and climatic requirements for the successful cultivation of tropical pumpkin (*Cucurbita moschata*), horned melon (*Cucumis metuliferus*), and bottle gourd (*Lagenaria siceraria*) are similar in that all three crops thrive in tropical and subtropical regions. Tropical pumpkin performs best in warm climates with optimal temperatures ranging between 20°C and 30°C. Growth is inhibited below 10°C, while excessive heat above 35°C can lead to flower drop and fruit deformities. The crop requires fertile, well-drained sandy loam or loamy soils with a pH of 6.0 to 7.0, enriched with organic matter such as compost or well-rotted manure to boost soil fertility and support vigorous growth. Regular moisture is essential, particularly during germination and fruit development, with an estimated requirement of 25 to 35 mm of water weekly. Drip irrigation is often recommended to supply consistent moisture while avoiding waterlogging and root rot [16, 17].

Horned melon shares similar thermal preferences, ideally between 20°C and 30°C, but it demonstrates better tolerance to dry and mildly cool environments, making it suitable for semiarid regions. It requires fertile, well-drained soils with high organic content and a pH of 6.0 to 6.8. Though relatively drought-resistant, horned melon responds well to regular irrigation during flowering and fruit setting, provided overwatering is avoided to prevent root diseases. Its vining nature benefits from trellising, which improves air circulation and helps reduce fungal infections [18]. Bottle gourd, on the other hand, prefers slightly higher temperatures ranging from 25°C to 35°C and grows well in loamy or sandy loam soils that are nutrient-rich and well-drained, with a pH of 6.0 to 7.0. It is moderately sensitive to drought, especially during flowering and fruiting stages, and requires approximately 20 to 30 mm of water per week. Drip irrigation is also suitable for maintaining optimal moisture levels without causing water stress or waterlogging [15, 19].

1.2.4 Crop agronomy

The agronomic practices for tropical pumpkin (*Cucurbita moschata*), horned melon (*Cucumis metuliferus*), and bottle gourd (*Lagenaria siceraria*) cover essential elements such as cultivation techniques, growth stages, pest control, and harvesting methods. Tropical pumpkin is usually planted during the warm growing season in well-prepared soils rich in organic matter, with recommended spacing of about 1 meter between plants to allow proper aeration and sunlight penetration. Horned melon is either directly sown after the last frost or started in seed trays, requiring 30 to 60 cm spacing between plants and well-drained fertile soils similar to those favored by cucumbers [18]. Bottle gourd is grown during warm seasons, with seeds sown either directly in the field or in nursery beds, spaced 1 to 2 meters apart depending on variety and management system [15]. All three crops require adequate sunlight and nutrient-rich soils for healthy development.

The crops follow similar developmental stages: seedling, vegetative growth, flowering, fruiting, and ripening. Tropical pumpkin reaches maturity within 80 to 100 days after planting, while horned melon matures in about 70 to 90 days and bottle gourd in 60 to 90 days [19, 20]. Pollination, mostly aided by bees and other insects, is critical for fruit formation. Pest management relies on integrated pest management (IPM) practices, including crop rotation, resistant varieties, regular field scouting, and the use of organic pesticides. Tropical pumpkin and bottle gourd are commonly affected by aphids, whiteflies, and fruit flies, while horned melon faces threats from aphids and spider mites. Organic mulching and intercropping are often used to suppress weeds and discourage pest buildup. Harvesting is done once fruits reach physiological maturity—indicated by changes in color and firmness—with careful handling using sharp tools to prevent bruising and preserve quality [18, 19].

Purpose: The purpose of this study was to identify high-performing accessions suitable for breeding, regional adaptation, and commercialization while also examining local consumption trends and market dynamics.

2. Materials and methods

2.1 Description of the study site

The study was conducted at the University of Nairobi Upper Kabete Field Station in Kabete Sub-county, Nairobi County, Kenya, approximately 11 km northwest of Nairobi city center at an altitude of 1940 meters above sea level [21]. The site is characterized by a bimodal rainfall pattern, with annual precipitation ranging from 900 to 1200 mm, long rains occurring from March to May, and short rains between October and December [22]. The mean annual temperature is around 18.7°C, with average maximum and minimum temperatures of 24.5°C and 12.8°C respectively. Potential evaporation averages 1500 mm annually, while actual evapotranspiration is approximately 1000 mm, depending on the season and crop cover. The soils at Upper Kabete are primarily deep, well-drained humic nitisols with a friable clay texture and good fertility, derived from volcanic parent material, making them suitable for diverse crop cultivation [16, 17]. This site serves as a key center for agricultural research and training and offers favorable conditions for cultivating underutilized cucurbits such as tropical pumpkin (*Cucurbita moschata*), horned melon (*Cucumis metuliferus*), and bottle gourd (*Lagenaria siceraria*), supporting their evaluation for growth performance and agronomic traits under managed field conditions.

2.2 Plant material and seed preparation

Fifteen accessions of underutilized cucurbit crops—tropical pumpkin (*Cucurbita moschata*), horned melon (*Cucumis metuliferus*), and bottle gourd (*Lagenaria siceraria*)—were selected for this study based on agroecological diversity, availability, and genetic improvement potential. These accessions were sourced from farmers' fields, wild populations, and research institutions across Kenya. For tropical pumpkin, five accessions included local landraces from Western, Eastern, and Central Kenya (KE-PMK-001 to KE-PMK-015), smallholder farms in Embu (Embu-PMK-01), Meru (Meru-PMK-03), KALRO stations (KALRO-PMK-A), and the University of Nairobi (UoN-PMK-05). Horned melon accessions comprised KE-HML-001 and KE-HML-002 (from Kitui and Embu), Kiboko-HML-02 (KALRO), Tharaka-HML-03 (dryland-adapted from Tharaka-Nithi), and Kakuma-HML-01 (a wild type from Turkana). Bottle gourd accessions included KE-BLG-001 and KE-BLG-002 (from Nyanza and Western Kenya), Kisii-BLG-01 (large-fruited), Makueni-BLG-02 (early maturing), and Meru-BLG-05 (used in traditional cooking). Mature fruits were harvested at full ripeness, and seeds were extracted using sterile tools, fermented in clean containers for 3 days to remove mucilage and enhance germination [10], then washed, sun-dried, and winnowed [6]. The seeds were stored under cool, dry conditions until planting, providing a genetically diverse base for evaluating morphological, agronomic, and yield-related traits.

2.3 Experimental design, layout, and market survey

The study aimed to evaluate the genetic diversity, agronomic performance, and market potential of tropical pumpkin (*Cucurbita moschata*), horned melon (*Cucumis metuliferus*), and bottle gourd (*Lagenaria siceraria*) in selected regions of Kenya, using a combination of field experimentation and market surveys to provide comprehensive insights into both production and commercialization. The field component was conducted at the University of Nairobi Research Farm using a randomized complete block design with three replications. Fifteen accessions—five per species—were selected based on agroecological diversity and availability, including KE-PMK-001 to UON-PMK-05 for tropical pumpkin, KE-HML-001 to Kakuma-HML-01 for horned melon, and KE-BLG-001 to Meru-BLG-05 for bottle gourd. Each accession was planted in 3 × 3 meters plots, with 1-meter alleys to prevent interference, and seeds were spaced at 1 × 1 meter, later thinned to one seedling per hill. Before planting, 30 tons per hectare of well-decomposed farmyard manure and 200 kg per hectare of Triple Super Phosphate were applied. Embu-PMK-01 was planted as a border crop to minimize edge effects. All recommended agronomic practices were applied uniformly, and trials ran across two cropping seasons—October 2023 to January 2024, and March to July 2024—to assess performance under different conditions. The market survey was carried out in Nairobi, Meru, Kisumu, and Kakamega using structured questionnaires and direct interviews with traders and consumers in major markets, including Wakulima, Makutano, Kibuye, and Kakamega Town markets. The survey explored crop availability, supply trends, pricing, consumer preferences, and awareness levels, particularly for horned melon, which remains less familiar in many parts of Kenya.

2.4 Data collection

This study combined field evaluation and a market survey to assess the genetic diversity, agronomic performance, and market potential of tropical pumpkin (*Cucurbita moschata*), horned melon (*Cucumis metuliferus*), and bottle gourd (*Lagenaria siceraria*) in Kenya. Fifteen accessions—five per crop—were evaluated at the University of Nairobi Research Farm using a randomized complete block design with three replications. Morphological and agronomic traits were assessed using a modified descriptor list adapted from the International Plant Genetic Resources Institute (IPGRI), with data collected on qualitative traits such as leaf shape, vine growth habit, and fruit and seed characteristics, as well as quantitative traits including vine length, number of fruits per plant, fruit weight, seed size, and days to flowering and maturity. Observations were made from four tagged plants per plot, and precision instruments such as digital calipers and electronic balances were used for accurate measurement. Complementing the field study, a market survey was carried out in Nairobi, Meru, Kisumu, and Kakamega through structured questionnaires and interviews with traders, vendors, and consumers in local markets. The survey aimed to identify preferred fruit traits, sourcing patterns, pricing trends, and consumer awareness, with special attention to horned melon, which remains underutilized and less familiar among Kenyan consumers.

2.5 Data analysis

The study combined both field experiments and market surveys to assess the genetic diversity, agronomic performance, and market potential of tropical pumpkin, horned melon, and bottle gourd. Data were collected over two growing seasons and analyzed using two-way Analysis of Variance (ANOVA) to detect significant differences among the 15 accessions, seasonal variations, and their interactions at a 5% significance level. Where significant differences were found, means were separated using the Student's Newman-Keuls test. Pearson correlation analysis was also conducted to explore the relationships among key traits such as vine length, fruit number, and fruit weight.

Both qualitative and quantitative trait data were organized into a matrix for cluster analysis. Genetic dissimilarity among accessions was computed using Euclidean distance, and a dendrogram was generated through the Unweighted Pair Group Method with Arithmetic Mean (UPGMA). Principal Component Analysis (PCA) was also performed to identify the traits that contributed most to the variation among accessions. On the market side, data from four major towns—Nairobi, Meru, Kisumu, and Kakamega—were summarized using descriptive statistics. This included frequencies and percentages to illustrate crop availability, sourcing patterns, price variations, and consumer preferences. Regional comparisons highlighted the most valued traits across different markets. These findings were then integrated with field performance data to provide insights that could guide future breeding programs and support the promotion of improved cucurbit varieties.

3. Results

3.1 Assessment of qualitative and quantitative traits

The results revealed significant differences ($p \leq 0.05$) among the 15 accessions of tropical pumpkin, horned melon, and bottle gourd across all evaluated quantitative and agronomic traits. Main vine length ranged from 240.83 cm in Makueni-BLG-02 to 273.58 cm in Tharaka-HML-03, with Tharaka-HML-03, Embu-PMK-01, and Meru-BLG-05 exhibiting the longest vines, while Makueni-BLG-02 and Kiboko-HML-02 had the shortest. Branch number varied significantly, with Kisii-BLG-01 recording the highest (16.25), followed by Tharaka-HML-03 (16.00) and the lowest values observed in Kakuma-HML-01 (13.58) and KE-PMK-003 (13.92). Fruit number per plant also differed notably, with Tharaka-HML-03 producing the most (15.60), followed by Kisii-BLG-01 (14.00) and Meru-PMK-03 (14.12), while Kiboko-HML-02 (8.80) and Makueni-BLG-02 (9.56) recorded the fewest. Average fruit weight ranged from 198.67 g in KE-BLG-001 to 315.67 g in Meru-PMK-03, with Meru-PMK-03 and UoN-PMK-05 yielding the heaviest pumpkin fruits, and Kiboko-HML-02 and Makueni-BLG-02 having the lightest horned melon and bottle gourd fruits, respectively. Seed size also varied, from 3.30 mm in Makueni-BLG-02 to 4.18 mm in Tharaka-HML-03, with Tharaka-HML-03 and UoN-PMK-05 having the largest seeds. Rind thickness ranged between 3.16 mm in Makueni-BLG-02 and 4.33 mm in Tharaka-HML-03, with the thickest rinds observed in Tharaka-HML-03 and Meru-PMK-03, and the thinnest in Makueni-BLG-02 and Kiboko-HML-02. These findings confirm considerable genetic variability among the accessions, which is useful for selection and crop improvement (**Table 2**).

The cluster dendrograms illustrate the morphological relationships among accessions of three cucurbit crops—tropical pumpkin, horned melon, and bottle gourd—based on their trait similarities. In the tropical pumpkin group, accessions such as EF/MM1-01, AIW/AIHC-03, and Poohei-03 formed a tight cluster, suggesting high morphological similarity, while ThromMel-03 appeared more distinct. Similarly, in the Horned Melon group, EF/MM1-01 and KuroConMel-01 were closely linked,

Crop and accession	Main vine length (cm)	Branch number	Fruit number	Fruit weight (g)	Seed size (mm)	Rind thickness (mm)
Tropical pumpkin						
Embu-PMK-01	270.33abc	15.83abcd	12.87cde	298.00abc	3.90cde	3.84de
Meru-PMK-03	262.17abcd	14.50cde	14.12abc	315.67ab	4.05abc	4.18ab
KE-PMK-001	255.42def	15.08bcd	10.73efg	284.50bcd	3.66f	3.72ef
KE-PMK-003	248.00fgh	13.92ef	9.85fg	275.33cd	3.58f	3.40h
UoN-PMK-05	266.25abcd	15.17bcd	11.21def	301.67abc	4.12ab	4.05bc
Horned melon						
KE-HML-001	259.67cde	14.66de	13.34bcd	208.33f	3.78def	3.88cd
Tharaka-HML-03	273.58ab	16.00ab	15.60a	224.00ef	4.18a	4.33a
Kiboko-HML-02	245.92h	15.33bc	8.80g	219.67ef	3.50f	3.29i
Kakuma-HML-01	251.17efg	13.58f	10.95efg	232.33de	3.85cde	3.54gh
KEPHIS-HML-05	258.42cde	14.25ef	12.56cde	228.00ef	3.91cde	3.86cd
Bottle gourd						
KE-BLG-001	247.75fgh	15.42bc	11.88def	198.67f	3.61f	3.40h
Kisii-BLG-01	258.00cde	16.25a	14.00abcd	210.50f	3.97bcd	3.82de
Makueni-BLG-02	240.83h	14.17ef	9.56fg	202.00f	3.30g	3.16j
Meru-BLG-05	264.42abcd	14.83cde	12.10def	215.00f	4.00abc	3.90cd
UoN-BLG-06	261.25abcd	15.58abc	13.44abc	218.00ef	3.85cde	3.71ef
P-value	0.031	0.015	0.002	0.007	<0.0001	<0.0001
SE	8.57	0.84	1.17	11.93	0.06	0.11

Means followed by the same letter in a column are not significantly different at $p \leq 0.05$ using Student's Newman-Keuls (SNK) test. SE = standard error.

Table 2.
Variation in quantitative and agronomic traits among cucurbit accessions.

indicating similar traits, whereas ThomMel-03 stood out as the most divergent. For bottle gourd, IAMFo-03 and MAI-001 exhibited close similarity, with EF/BO02-01 and BottleBiG-01 showing moderate to lower similarity. The vertical distances, which represent dissimilarity values, indicate that shorter linkages reflect closer morphological relationships (**Figure 4**).

3.2 Assessment of agronomic traits

Significant variation ($p < 0.0001$) was observed across all agronomic traits among the accessions of tropical pumpkin, horned melon, and bottle gourd. In tropical pumpkin, germination ranged from 89.33% (KE-PMK-003) to 96.67% (Embu-PMK-01), with emergence occurring between 8.83 and 9.67 days. Embu-PMK-01 flowered and matured earliest, while KE-PMK-003 was latest. Horned melon accessions showed germination between 90.17% (Kakuma-HML-01) and 97.83% (KE-HML-001), with emergence, flowering, and maturity also showing significant variation—Kakuma-HML-01 being earliest and Kiboko-HML-02 latest. For bottle gourd, germination ranged from 91.33% to 96.00%, with UoN-BLG-06 emerging, flowering, and maturing earliest, while Makueni-BLG-02 was the latest. These findings highlight substantial genetic diversity in early growth and reproductive timing among the accessions (**Table 3**).

3.3 Principal component analysis (PCA)

The scree plot illustrates the distribution of eigenvalues and cumulative variability for tropical pumpkin, horned melon, and bottle gourd accessions. In all three crops, the first few principal components account for most of the variability, with a steep decline in eigenvalues observed from component 1 to 5. Tropical pumpkin and bottle gourd recorded higher initial eigenvalues compared to horned melon. The cumulative variability approached saturation near component 10 in all crops, with over 90% of total variation explained by the first 7 to 10 components. This indicates that a few principal components capture the majority of variation among accessions for each crop (**Figure 5**).

3.4 Contribution of variables to observed variation in three cucurbit crops

The results in **Table 3** show that across all three cucurbit crops—tropical pumpkin, horned melon, and bottle gourd—the first principal component (F1) accounted

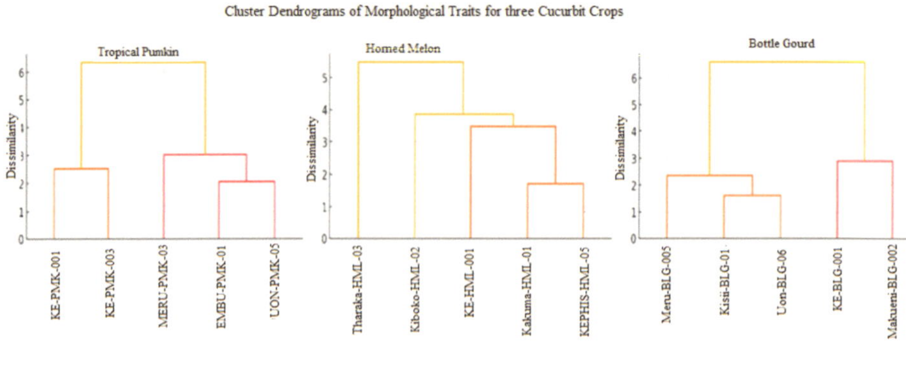

Figure 4.
Cluster dendrogram depicting morphological traits for three cucurbit crops.

Crop	Accessions	Percent germination	Days to emergence	Days to male flowering	Days to female flowering	Days to maturity
Tropical pumpkin	Embu-PMK-01	96.67a	9.00bc	62.50cd	64.00cd	109.33cd
	Meru-PMK-03	93.00ab	9.50ab	64.17bc	66.67bc	112.00bc
	KE-PMK-001	91.50ab	9.67a	65.33ab	67.83ab	114.17ab
	KE-PMK-003	89.33b	9.17bc	66.83a	69.00a	116.50a
	UoN-PMK-05	92.67ab	8.83c	63.00cd	65.33cd	110.83c
	P value	<0.0001	<0.0001	<0.0001	<0.0001	<0.0001
	SE	3.984	0.187	0.877	0.854	1.143
Horned melon	KE-HML-001	97.83a	9.33ab	64.67ab	66.00ab	111.00b
	Tharaka-HML-03	95.33ab	9.17ab	63.50bc	64.67bc	109.50bc
	Kiboko-HML-02	92.50bc	9.50a	65.83a	67.50a	113.67a
	Kakuma-HML-01	90.17c	9.00b	62.00c	63.00c	107.33c
	KEPHIS-HML-05	94.67ab	8.83b	63.33bc	65.00bc	110.00bc
	P value	<0.0001	<0.0001	<0.0001	<0.0001	<0.0001
	SE	3.472	0.198	0.821	0.744	1.216
Bottle gourd	KE-BLG-001	96.00a	8.83b	62.83bc	64.33bc	109.67c
	Kisii-BLG-01	94.17ab	9.17ab	64.50ab	66.17ab	112.00b
	Makueni-BLG-02	91.33bc	9.33a	65.67a	67.83a	114.83a
	Meru-BLG-05	92.67bc	9.00ab	63.17bc	64.83bc	110.50bc
	UoN-BLG-06	93.83ab	8.67b	61.67c	63.17c	108.33c
	P value	<0.0001	<0.0001	<0.0001	<0.0001	<0.0001
	SE	3.812	0.176	0.789	0.799	1.184

Means followed by the same letter within a column are not significantly different based on SNK test at $p \leq 0.05$.
SE = standard error.

Table 3.
Variation in agronomic traits of selected cucurbit accessions.

for the highest proportion of total variability, with eigenvalues exceeding 410 and individual variation ranging from 70.63% to 71.21%. The primary contributors to F1 were fruit weight and main vine length, indicating their strong influence on overall agro-morphological variation. The second component (F2) contributed an additional 15.38% to 15.70% of variability and was mainly associated with days to maturity and flowering traits. The third component (F3) was captured between 10.48% and 10.72% of the total variation, primarily influenced by days to male and female flowering and days to maturity. Collectively, the three components explained over 96% of the total variability in each crop, demonstrating a high level of explained variation by the selected agro-morphological traits (**Table 4**).

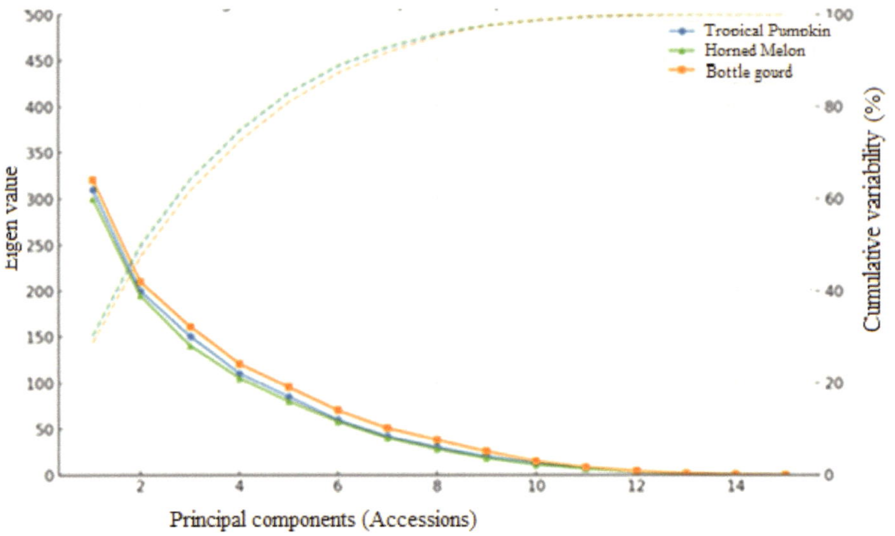

Figure 5.
Scree plot of Eigen values for tropical pumpkin, horned melon, and bottle gourd in Kenya.

Agro-morphological variables	F1 (P)	F1 (M)	F1 (G)	F2 (P)	F2 (M)	F2 (G)	F3 (P)	F3 (M)	F3 (G)
Main vine length	0.310	0.298	0.315	94.21	95.50	95.85	3.220	3.010	3.150
Branch number	0.003	0.005	0.006	0.120	0.115	0.110	0.125	0.120	0.105
Rind color	0.000	0.000	0.000	0.055	0.048	0.050	0.270	0.265	0.260
Fruit shape	0.018	0.016	0.014	0.040	0.045	0.038	0.260	0.255	0.250
Seed shape	0.002	0.003	0.001	0.105	0.115	0.110	0.006	0.007	0.004
Fruit number	0.009	0.007	0.010	0.042	0.048	0.050	0.102	0.101	0.095
Days to seed emergence	0.017	0.021	0.018	0.011	0.013	0.009	0.088	0.085	0.087
Days to male flowering	0.048	0.040	0.043	0.180	0.170	0.160	21.90	22.10	22.00
Days to female flowering	0.005	0.007	0.004	0.190	0.185	0.195	17.60	17.80	17.700
Days to maturity	0.022	0.025	0.018	3.010	2.800	2.700	54.70	55.30	54.90
Fruit weight	98.95	99.10	99.20	0.290	0.320	0.300	0.014	0.016	0.012
Seed size	0.004	0.006	0.003	0.008	0.012	0.009	0.017	0.014	0.016
Rind thickness	0.000	0.001	0.000	0.080	0.070	0.078	0.030	0.029	0.027
Eigenvalue	412.8	410.3	414.12	89.45	91.20	90.65	62.30	61.75	60.98
Variability (%)	70.97	70.63	71.21	15.38	15.70	15.59	10.72	10.63	10.48
Cumulative %	70.97	70.63	71.21	86.35	86.33	86.79	97.07	96.97	97.27

Legend: F1–F3: *First three principal components (variability factors),* **P**: *tropical pumpkin,* **M**: *horned melon, and* **G**: *bottle gourd.*

Table 4.
Contribution of variables to observed variation in three cucurbit crops.

3.5 Market survey

The study revealed significant differences in market source, availability, and pricing of tropical pumpkin, horned melon, and bottle gourd across Nairobi, Meru, Kisumu, and Kakamega. Tropical pumpkin was the most widely available, sourced locally in all markets, with availability ranging from 80% in Kisumu to 90% in Nairobi, and moderately priced between KSh 80 and 200 per kilogram. Bottle gourd had moderate availability (60–70%) depending on the region, with the lowest price range of KSh 40–90, particularly affordable in Meru. Horned melon, though highest in price (KSh 150–200), was the least available, with presence below 60% in all regions and mostly supplied through specialized urban sources. These market differences reflect variations in supply chains, production intensity, and consumer demand, with statistical tests confirming that the crops differed significantly in availability and pricing (**Table 5**, **Figure 6**).

The results on consumer preferences and trader priorities reveal distinct patterns for each cucurbit crop. Consumers preferred tropical pumpkin mainly for its taste and shelf life, while traders valued its long shelf life. For horned melon, unique flavor and health benefits were the most important traits to consumers, while traders focused on its attractive color. Bottle gourd was favored by consumers for its tender flesh and ease of cooking, with traders prioritizing uniform shape and size (**Table 6**).

Across the four markets, tropical pumpkin was the most available crop, while horned melon consistently had the highest price. Consumer preferences varied slightly by region—taste and health benefits were emphasized in Nairobi,

Crop	Market	Retail price (KSh/kg)	Availability (%)
Tropical pumpkin	Nairobi	150–200^{b}	90^{a}
Tropical pumpkin	Meru	80–120^{c}	85^{a}
Tropical pumpkin	Kisumu	100–150^{c}	80^{a}
Tropical pumpkin	Kakamega	90–130^{c}	85^{a}
Significance		a > b > c	a > b > c
Horned melon	Nairobi	150–200^{b}	50^{c}
Horned melon	Meru	100–140^{c}	45^{c}
Horned melon	Kisumu	130–160^{b}	55^{bc}
Horned melon	Kakamega	140–170^{b}	60^{bc}
Significance		a > b > c	a > b > c
Bottle gourd	Nairobi	60–90^{d}	65^{b}
Bottle gourd	Meru	40–60^{d}	70^{b}
Bottle gourd	Kisumu	50–80^{d}	60^{bc}
Bottle gourd	Kakamega	55–85^{d}	68^{b}
Significance		a > b > c > d	a > b > c

Superscript letters ($^{a, b, c, d}$) indicate significance groupings based on ANOVA and post hoc tests; values sharing the same letter are not significantly different ($p \leq 0.05$). Significance rows summarize relative groupings among entries per crop.

Table 5.
Retail prices, availability, and significance indicators for cucurbit crops.

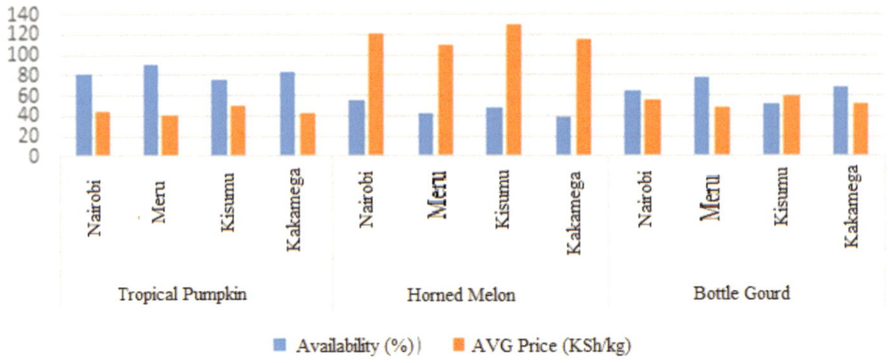

Figure 6.
Availability and average price of the three cucurbits.

Crop	Consumer preferred traits	Trader valued trait
Tropical pumpkin	Taste (78%), shelf life (62%)	Long shelf life (68%)
Horned melon	Unique flavor (65%), health benefits (59%)	Attractive color (72%)
Bottle gourd	Tender flesh (71%), ease of cooking (58%)	Uniform shape and size (61%)

Table 6.
Consumer preferences and traits valued by traders across all markets.

Market	Most available crop	Highest priced crop	Consumer priority	Trader challenge
Nairobi	Tropical pumpkin	Horned melon	Taste and health	Supply consistency
Meru	Tropical pumpkin	Horned melon	Taste and shelf life	Market glut
Kisumu	Tropical pumpkin	Horned melon	Nutrition and price	Storage loss
Kakamega	Tropical pumpkin	Horned melon	Tenderness	Price fluctuation

Table 7.
Comparative summary of market differences.

taste and shelf life in Meru, nutrition and price in Kisumu, and tenderness in Kakamega. Traders faced different challenges in each market, including supply consistency in Nairobi, market glut in Meru, storage losses in Kisumu, and price fluctuation in Kakamega. These findings highlight how consumer demand and trader experiences influence the marketing and promotion of cucurbit crops across different regions (**Table 7**).

4. Discussion

The evaluation of 15 accessions across tropical pumpkin, horned melon, and bottle gourd revealed significant morphological and agronomic diversity, particularly in traits such as vine length, branch number, fruit yield, fruit weight, seed size, and rind thickness ($p \leq 0.05$). Notable performers included Tharaka-HML-03, Embu-PMK-01, and Meru-BLG-05 for vine length, while Kisii-BLG-01 and Tharaka-HML-03 excelled in branching and fruit production. Heavier fruits were observed in Meru-PMK-03 and UoN-PMK-05, both pumpkin accessions, while horned melon and bottle gourd tended to have lighter fruits, supporting earlier findings by [23, 24] on cucurbit variability. Seed size and rind thickness also varied significantly, with Tharaka-HML-03 emerging as a strong candidate for breeding due to its consistent performance, echoing insights from [25]. Dendrogram analysis confirmed these patterns, with tropical pumpkin accessions showing both tight and distinct clusters, horned melon displaying moderate diversity, and bottle gourd forming compact groupings—patterns aligned with findings from [26, 27] on agro-morphological clustering in local cucurbits. This morphological variation is significant for breeding programs, as it enables the selection of elite traits suited to specific agroecological zones and market demands. The wide variability observed in traits like fruit weight and maturity duration confirms the adaptability and genetic richness within Kenya's cucurbit germplasm, as previously documented by [28] in participatory varietal selection studies across western Kenya.

The study revealed significant genetic and phenotypic diversity among the 15 cucurbit accessions, as reflected in key agronomic and morphological traits. Tropical pumpkin accessions exhibited notable differences in percent germination, flowering time, and maturity, with KE-PMK-003 maturing latest and Embu-PMK-01 maturing earliest, indicating differential adaptability—a finding consistent with [13], who highlighted maturity variability in Kenyan pumpkin landraces. Horned melon accessions also showed distinct variation, particularly in flowering and germination; Kakuma-HML-01 and Kiboko-HML-02 differed markedly in maturity, while KE-HML-001 recorded superior germination, echoing the adaptive diversity reported by [23] across Kenya's agroecological zones. Among bottle gourd accessions, KE-BLG-001 exhibited the highest germination rates, whereas UoN-BLG-06 was earliest to flower and mature—traits of agronomic value in regions facing weather variability or pest pressure. These findings align with observations by reference [27], who documented maturity variation in local bottle gourd germplasm, supporting the crop's suitability for selective breeding and region-specific cultivation.

Principal component analysis (PCA) served as a powerful tool for distinguishing trait contributions to variation among cucurbit accessions, a method widely employed in similar studies. The dominance of fruit weight and vine length in the first principal component mirrors the findings of [24], who reported fruit size as a key factor in cucurbit diversity. Likewise, flowering and maturity traits, which shaped the second and third components, have been consistently identified in previous analyses as central to adaptation and varietal selection [26, 29]. The application of PCA in this context aligns with approaches used by reference [28], who emphasized the method's value in isolating high-performing lines for breeding. Comparable multivariate studies, such as those by reference studies [23, 25], also underscored the importance of integrating morphological and temporal traits to guide selection decisions. Collectively, these studies support the relevance of PCA in identifying elite accessions and structuring breeding programs aimed at enhancing local cucurbit germplasm.

Market surveys across Kenyan regions have increasingly highlighted the complex interplay between crop availability, pricing, and consumer preferences, findings that are consistent with a growing body of literature on indigenous vegetables and underutilized crops. The observed regional variation in availability—particularly the dominance of tropical pumpkin and limited access to horned melon—reflects patterns documented by [30], who reported that market supply is closely tied to proximity of production zones and cultivation intensity. Similarly, disparities in pricing, with horned melon fetching premium rates due to its scarcity and perceived health value, align with findings by reference [31], who noted that consumer awareness and product novelty significantly shape pricing trends in fresh produce markets. Furthermore, the differentiated consumer and trader preferences across the three cucurbit crops reinforce trends observed by reference [12], where sensory traits (e.g., taste and texture) and visual appeal (e.g., color and uniformity) were central to product acceptance in urban markets. Collectively, these studies affirm that market potential and crop adoption strategies must consider both agronomic factors and end-user perceptions, particularly when promoting lesser-known crops like horned melon or positioning bottle gourd for cost-sensitive markets.

Comparative assessments of market challenges across regions have been widely acknowledged in agricultural market studies, reinforcing the view that crop marketing success is deeply influenced by localized dynamics. The supply inconsistency in urban areas like Nairobi and post-harvest losses in Kisumu concurs with the finding [30, 32], who emphasized that logistical constraints and infrastructural inadequacies often hinder smallholder access to stable markets. Similarly, the market glut reported in Meru is reflective of challenges reported in Ref. [31], where uncoordinated production and limited value chain integration led to periodic oversupply and waste. Concerns over price fluctuation in places like Kakamega have also been previously attributed to weak aggregation systems and fluctuating consumer demand, as noted in Refs. [21, 31] on staple and horticultural crops. These parallels affirm the need for region-specific interventions, including investment in storage infrastructure, structured marketing channels, and participatory breeding that incorporates both agronomic and market traits. Aligning crop improvement with market realities, as documented in Refs. [32, 33], enhances not only varietal adoption but also the overall sustainability and profitability of cucurbit cultivation in Kenya.

5. Conclusion

The study comprehensively evaluated 15 accessions of tropical pumpkin, horned melon, and bottle gourd across different Kenyan regions, revealing significant agro-morphological and market-based variability with strong potential for crop improvement and market integration. Field results showed notable diversity in traits such as vine length, fruit number, weight, and seed characteristics, with accessions like Tharaka-HML-03, Embu-PMK-01, and Meru-BLG-05 standing out for productivity and growth vigor. Variability in germination, flowering, and maturity across accessions—such as early emergence in Embu-PMK-01 and UoN-BLG-06, and high germination in KE-HML-001—highlighted adaptability to distinct agro-ecological zones [21, 25]. Principal component analysis indicated that a few traits, especially fruit weight and vine length, accounted for over 70% of phenotypic variability, simplifying selection targets for breeding [24, 28]. The market survey showed tropical pumpkin as the most available crop, largely locally sourced, while horned melon, though least available, fetched the highest prices due to its niche demand; bottle gourd was moderately available and most affordable, particularly in Meru. These patterns reflect previous findings that link availability and proximity to pricing and demand [30, 31]. Consumer preferences favored taste, tenderness, and shelf life, while traders prioritized visual traits and uniformity, with regional differences in priorities and challenges such as supply inconsistency and storage losses [32]. These findings align agronomic diversity with market demand, offering a practical framework for varietal selection and commercialization. The integration of field data with market insights provides a strong basis for developing resilient, high-yielding, and consumer-accepted cucurbit varieties suited to Kenya's diverse production systems.

6. Recommendations

Based on the identified gaps and areas for improvement, we make several recommendations to enhance cucurbit production and utilization in Kenya. These include conducting multilocation trials to validate the performance of promising accessions across diverse agroecological zones and strengthening postharvest research to reduce storage losses and improve the shelf life of perishable crops like horned melon and bottle gourd. Further, enhancing market linkage studies is crucial to address supply chain challenges and price instability, especially for underutilized crops. Consumer-oriented breeding should also be prioritized to incorporate preferred traits such as flavor, health benefits, tenderness, and ease of cooking of the new varieties. Lastly, there is a need to promote farmer training on improved production practices, seed selection, and value addition to increase yields and income for smallholder growers.

Acknowledgements

I wish to sincerely thank my fellow colleagues Profs. William M. Muiru and John W. Kimenju for their invaluable contribution and assistance. The University of Nairobi is also appreciated for providing facilities and resources that made this study a reality. All technical staff, farmers, and other stakeholders who participated in this research are highly appreciated. Artificial Intelligent (AI) assisted tools were used to improve the language and grammar of this publication to give it clarity. However, scientific content, data analysis and result interpretations were done by the Authors.

Conflict of interest

The authors declare no conflict of interest.

Appendix

Cucurbit market survey questionnaire

Section A: Respondent information

Question	Response
Name (optional)	______________
Gender	☐ Male ☐ Female
Age group	☐ 18–25 ☐ 26–35 ☐ 36–45 ☐ 46+
Market location	☐ Nairobi ☐ Meru ☐ Kisumu ☐ Kakamega
Type of respondent	☐ Trader ☐ Consumer
Number of years in trading cucurbits	________ years

Section B: Availability of cucurbit crops

B1. Which of the following crops do you sell or buy regularly? (Tick all that apply)

- ☐ Tropical pumpkin
- ☐ Horned melon
- ☐ Bottle gourd

B2. How often are these crops available in your market?

Crop	Always available	Sometimes available	Rarely available	Never available
Tropical pumpkin	☐	☐	☐	☐
Horned melon	☐	☐	☐	☐
Bottle gourd	☐	☐	☐	☐

Section C: Source of cucurbit crops

C1. Where do you source the following crops? (Tick all that apply)

Crop	Local farmers	Peri-urban farms	Urban markets	Importers
Tropical pumpkin	☐	☐	☐	☐
Horned melon	☐	☐	☐	☐
Bottle gourd	☐	☐	☐	☐

Section D: Pricing information

D1. What is the average price range for each crop in your market?

Crop	Lowest price (KSh/kg)	Highest price (KSh/kg)
Tropical pumpkin	________	________
Horned melon	________	________
Bottle gourd	________	________

Section E: Consumer preferences

E1. Which traits do your customers prefer most when purchasing cucurbits? (Tick top 2 per crop)

Trait	Tropical pumpkin	Horned melon	Bottle gourd
Taste	☐	☐	☐
Shelf life	☐	☐	☐
Unique flavor	☐	☐	☐
Ease of cooking	☐	☐	☐

Trait	Tropical pumpkin	Horned melon	Bottle gourd
Health benefits	☐	☐	☐
Attractive color	☐	☐	☐
Tender flesh	☐	☐	☐
Uniform shape/size	☐	☐	☐

Section F: Trader challenges and recommendations

F1. What are the main challenges you face when trading cucurbits?

☐ Price fluctuations
☐ Market glut
☐ Inconsistent supply
☐ Postharvest losses
☐ Low consumer demand
☐ Poor storage infrastructure
☐ Transport challenges
☐ Others: ____________________

F2. Suggest improvements that would help you trade cucurbits more effectively:

____
____
____
____

Author details

Samuel Linguya Kimaru*, William Maina Muiru and John Wangai Kimenju
Department of Plant Science and Crop Protection, University of Nairobi, Nairobi, Kenya

*Address all correspondence to: samuelklinguya@gmail.com

References

[1] Grubben GJH, Denton OA, editors. Plant Resources of Tropical Africa 2: Vegetables. Wageningen, The Netherlands: PROTA Foundation. Co-published with Backhuys Publishers and CTA; 2004

[2] Alemu TT, Intipunya P, Gebeyo BA. A comprehensive review of edible coatings for postharvest management of fruits and vegetables: Enhancing food and nutrition security. Discover Agriculture. 2025;**3**:Article 190. DOI: 10.1007/s44279-025-00348-8

[3] Scortichini M. Sustainable management of diseases in horticulture: Conventional and new options. Horticulturae. 2022;**8**(6):517. DOI: 10.3390/horticulturae 8060517

[4] Masika FB, Alicai T, Shimelis H. Pumpkin and watermelon production constraints and management practices in Uganda. CABI Agriculture and Bioscience. 2022;**3**:Article 39. DOI: 10.1186/s43170-022-00101-x

[5] Maluleke MK, Moja SJ, Nyathi M, Modise DM, Mabhaudhi T. Nutrient concentration of African horned cucumber (*Cucumis metuliferus* L.) fruit under different soil types, environments, and varying irrigation water levels. Horticulturae. 2021;**7**(4):76. DOI: 10.3390/horticulturae7040076

[6] Aliero AA, Gumi AM. Studies on the germination, chemical composition and antimicrobial properties of *Cucumis metuliferus*. Annals of Biological Research. 2012;**3**(8):4059-4064

[7] Šeregelj V, Šovljanski O, Tumbas Šaponjac V, Vulić J, Ćetković G, Markov S, et al. Horned melon (*Cucumis metuliferus* E. Meyer Ex. Naudin)—Current knowledge on its phytochemicals, biological benefits, and potential applications. Processes. 2022;**10**(1):94. DOI: 10.3390/pr10010094

[8] Lim TK. Edible Medicinal and Non-Medicinal Plants: Volume 2, Fruits. Springer Science + Business Media; 2012. DOI: 10.1007/978-94-007-1764-0

[9] Cantwell MI. Postharvest Handling and Utilization of Cucurbits (UC Postharvest Research & Extension Center Report). Davis: University of California; 2011

[10] Olarewaju OO, Fajinmi OO, Arthur GD, Naidoo K. Food and medicinal relevance of *Cucurbitaceae* species in eastern and southern Africa. Bulletin of the National Research Centre. 2021;**45**:208. DOI: 10.1186/s42269-021-00659-y

[11] Achigan-Dako EG, Vodouhè RS, Sangare A. Characterization of local varieties of *Cucumis sativus* L. in Benin. African Journal of Agricultural Research. 2008;**3**(1):1-8

[12] Dembitsky VM, Poovarodom S, Leontowicz H, et al. The multiple nutrition properties of some exotic fruits: Biological activity and active metabolites. Food Research International. 2011;**44**(7):1671-1701. DOI: 10.1016/j.foodres.2011.03.003

[13] Chivenge P, Mabhaudhi T, Modi AT, Mafongoya P. The potential role of neglected and underutilised crop species as future crops under water-scarce conditions. International Journal of Environmental Research and Public Health. 2015;**12**(6):5685-5711. DOI: 10.3390/ijerph120605685

[14] Saltveit ME. Respiratory metabolism. In: Yahia EM, editor. Postharvest Physiology and Biochemistry of Fruits and Vegetables. Woodhead Publishing; 2019. pp. 73-91. DOI: 10.1016/B978-0-12-813278-4.00004-

[15] Mliki A, Staub JE, Sun Z, Ghorbel A. Genetic diversity in melon (*Cucumis melo* L.): An evaluation of African germplasm. Genetic Resources and Crop Evolution. 2001;**48**:587-597. DOI: 10.1023/A:1012598707354

[16] Phumzile Mkhize J, Mashilo J, Shimelis H. Progress on genetic improvement and analysis of bottle gourd (*Lagenaria siceraria*) for agronomic traits, nutrient composition, and stress tolerance. Frontiers in Sustainable Food Systems. 2021;**5**:Article 683635. DOI: 10.3389/fsufs.2021.683635

[17] Food and Agriculture Organization of the United Nations [FAO], International Soil Reference and Information Centre [ISRIC], & International Union of Soil Sciences [IUSS]. World Reference Base for Soil Resources 2011 (World Soil Resources Reports No. 106). FAO; 2011. ISBN 978-92-5-106765-7

[18] Food and Agriculture Organization of the United Nations (FAO). FAOSTAT: Crop Production and Yield Statistics for Pumpkins, Squash, and Gourds. FAO; 2019. Available from: https://www.fao.org/faostat/en/#data/QCL [Accessed: January 21, 2026]

[19] Owino MH, Gichimu BM, Muturi PW. Agro-morphological characterization of horned melon (*Cucumis metuliferus*) accessions from selected agro-ecological zones in Kenya. Australian Journal of Crop Science. 2020;**14**(9):1487-1496. DOI: 10.21475/ajcs.20.14.09.p2642

[20] Georges YKA, Sévérin BB, Nestor GBB, Kevin KK, Sélastique AD, Arsène ZBI. Agro-morphological variability of bottle gourd (*Lagenaria siceraria* (Molina) Standl.) accessions in Côte d'Ivoire. International Journal of Advances in Scientific Research and Engineering. 2019;**5**(7):129-143. DOI: 10.31695/IJASRE.2019.33416

[21] Nakazibwe IM, Wangalwa R, Olet EA, Kagoro GR. Local knowledge of pumpkin production, performance, and utilization systems for value addition avenues from selected agro-ecological zones of Uganda. African Journal of Agricultural Research. 2019;**14**(32):15091519. DOI: 10.5897/AJAR2019.14070

[22] Jaetzold R, Schmidt H, Hornetz B, Shisanya CA. Farm Management Handbook of Kenya: Volume II – Natural Conditions and Farm Management Information. 2nd ed. Nairobi, Kenya: Ministry of Agriculture; 2006

[23] Ondiek RA, Saber M, Abdel Fattah M. Spatial–temporal analysis of impacts of climate variability on maize yield in Kenya. Agriculture. 2024;**14**(8):1415. DOI: 10.3390/agriculture14081415

[24] Nyabera LA, Nzuki IW, Runo S, Amwayi PW. Assessment of genetic diversity of pumpkins (*Cucurbita* spp.) from western Kenya using SSR molecular markers. Molecular Biology Reports. 2021;**48**(1):1-12. DOI: 10.1007/s11033-021-06245-9

[25] Wamatu JN, Omollo EO, Onyango CM. Evaluation of yield and yield components of pumpkin landraces in Kenya. Kenya Agricultural Research Journal. 2012;**7**(2):102-108

[26] Karanja JK, Mugendi BJ, Khamis FM, Muchugi A. Nutritional evaluation of

some Kenyan pumpkins (*Cucurbita* spp.). International Journal of Agriculture and Forestry. 2014;**4**(3):195-200. DOI: 10.5923/j.ijaf.20140403.08

[27] Tanui M, Binott JJ, Dangasuk GO, Kipsumbai PK. Morphological and yield characteristics diversity among pumpkin accessions in Kenya. Journal of Crops, Livestock and Pest Management. 2023;**1**(2):29-48

[28] Morimoto Y, Gikungu M, Maundu P. Pollinators of the bottle gourd (*Lagenaria siceraria*) observed in Kenya. International Journal of Tropical Insect Science. 2004;**24**(1):79-86. DOI: 10.1079/IJT20046

[29] Muthoni J, Shimelis H. Pumpkin (*Cucurbita* ssp.): A neglected and underutilized crop with potential for production in Africa. Australian Journal of Crop Science. 2025;**19**(2):152-160. DOI: 10.21475/ajcs.25.19.02.p212

[30] Wadiwa J, Dongre H, Singh KV. Impact of integrated nutrient management on vegetative characteristics and yield of bottle gourd [*Lagenaria siceraria* (Mol.) Standl.] cv. Pusa Naveen. International Journal of Plant and Soil Science. 2025;**37**(5):304-312. DOI: 10.9734/ijpss/2025/v37i55454

[31] Odendo M, Ndinya C, Onyango E, Wanyama J, Akollo S, Makete N, et al. Does intervention in African indigenous vegetables value chain improve production and welfare outcomes? Evidence from western Kenya. Journal of Agriculture, Science and Technology. 2023;**22**(3):23-38. DOI: 10.4314/jagst.v22i3.3

[32] Gido EO, Ayuya OI, Owuor G, Bokelmann W. Consumption intensity of leafy African indigenous vegetables: Towards enhancing nutritional security in rural and urban dwellers in Kenya. Agricultural and Food Economics. 2017;**5**:14. DOI: 10.1186/s40100-017-0082-0

[33] Mwema C, Crewett W. Commercialization and market information constraints in African indigenous vegetables. Cogent Economics & Finance. 2019;**7**:1642173. DOI: 10.1080/23322039.2019.1642173

Chapter 4

In Vitro Applications in Cucumber (*Cucumis sativus* L.)

Sumeyra Ucar, Esma Yigider, Esra Yaprak, Ayse Gul Kasapoglu, Melek Ekinci, Ertan Yildirim and Murat Aydin

Abstract

Cucumber (*Cucumis sativus* L.), a vital horticultural crop in the Cucurbitaceae family, has been widely studied for its physiological, developmental, and genetic traits. *In vitro* culture techniques have played a critical role in facilitating genetic improvement, trait selection, and propagation of elite lines. This review comprehensively discusses the major *in vitro* techniques applied in cucumber research, including callus induction, shoot organogenesis, somatic embryogenesis, haploid production, and protoplast culture. Among various explants used, cotyledons, hypocotyls, and shoot apices are the most responsive for regeneration. Optimization of culture media especially the use of cytokinins (BAP, kinetin) and auxins (NAA, IAA) has significantly enhanced regeneration rates. In recent years, genetic transformation using *Agrobacterium tumefaciens* has gained prominence, particularly when coupled with *in vitro* regeneration systems. Furthermore, CRISPR/Cas9 genome editing has revolutionized cucumber biotechnology by enabling precise modifications for disease resistance, abiotic stress tolerance, and metabolic enhancements. Applications targeting genes such as MLO, eIF4E, and CsSGR have demonstrated remarkable potential in generating resistant and stress-tolerant cultivars. Hairy root systems have emerged as rapid platforms for functional gene analysis, especially for root-related traits. This chapter highlights the advancements and challenges in utilizing tissue culture and genome-editing techniques for improving cucumbers. Continuous innovations in explant selection, transformation efficiency, and regeneration protocols are essential to fully harness the crop's genetic potential.

Keywords: explant, micropropagation, organogenesis, plant regeneration, somatic embryogenesis

1. Introduction

Cucumber (*Cucumis sativus* L.) (2n = 2x = 14), which is a member of the diverse Cucurbitaceae family, is a widely cultivated vegetable crop with significant economic value worldwide [1]. It yields nutritious fruits rich in proteins, minerals, soluble carbohydrates, and vitamins. The fruit's polysaccharides and flavonoids are essential for antioxidant functions, cancer prevention, and managing liver inflammation [2]. A range of biotic and abiotic stresses often impact cucumber production. Significant challenges include insect pests and diseases, as well as nutrient deficiencies and adverse weather conditions, such as drought or extreme temperatures. These stresses can significantly reduce cucumber yields and even deteriorate fruit quality [3]. Cucumber cultivation suffers significant losses due to these challenges—some research suggests that combined biotic and abiotic stresses can lead to yield reductions of up to 80–97% in severe situations [3]. Traditional breeding approaches, such as crossbreeding (hybridization between distinct cucumber lines) and mutation breeding (inducing heritable changes through physical or chemical mutagens), have served as essential tools in cucumber (*Cucumis sativus* L.) improvement programs. Over recent decades, these conventional methods have significantly contributed to the development of new cultivars characterized by enhanced yield, fruit quality, and market adaptability. Traditional genetic breeding has laid the groundwork for many of the high-yielding, superior-quality cucumber varieties cultivated globally today [4]. However, despite these achievements, the advancement of traditional breeding remains inherently slow, often requiring multiple generations to stabilize desirable agronomic traits. A major limitation in cucumber breeding is its inherently narrow genetic base. Due to historical domestication bottlenecks, the species exhibits limited genetic diversity and a low level of natural variation within its germplasm [5]. This restricted genetic pool poses challenges in achieving sustained genetic gains and limits the breeder's ability to incorporate novel traits. Consequently, breeding programs may encounter a plateau in improvement unless innovative strategies—such as the integration of wild relatives, molecular tools, or tissue culture techniques—are employed to broaden genetic variation or overcome inherent limitations in the cucumber gene pool [6].

Plant tissue culture has emerged as a powerful adjunct to cucumber breeding, helping to address the above constraints in several ways. First, micropropagation techniques enable rapid clonal multiplication of desired cucumber genotypes, producing large numbers of identical plants in a short time frame and space, irrespective of season [7]. This is particularly valuable for propagating elite F_1 hybrids or genetically superior lines that cannot be easily maintained by seed due to segregation. Studies have shown that cucumber plants regenerated *in vitro* can retain genetic fidelity to the mother plant. For instance, regenerants derived via somatic embryogenesis have shown no significant molecular differences from the original F_1 hybrid, confirming the maintenance of genetic stability through the tissue culture process [8]. Such clonal propagation ensures that breeders can rapidly produce uniform planting material (disease-free and true-to-type), thereby overcoming the slow generation turnover of conventional breeding.

Tissue culture forms the basis for modern biotechnological techniques. An effective *in vitro* regeneration system is essential for genetic transformation in cucumber—introducing new genes, such as those for virus resistance or stress tolerance, through *Agrobacterium*-mediated transfer, depends on the ability to regenerate whole plants from transformed cells [9]. In practice, cucumber's genotype dependence in regeneration posed a challenge, but optimized tissue culture protocols (utilizing responsive cultivars like Poinsett 76 or 9930) have allowed researchers to produce transgenic cucumber plants routinely. This capability enables precise genetic enhancements that surpass the limitations of a limited gene pool, allowing the introduction of new traits or gene edits that crossing alone cannot achieve [10].

In vitro selection techniques, combined with tissue culture, can help develop stress-tolerant lines by exposing cell cultures or regenerating plantlets to selective agents such as high salinity or pathogens [11]. This process enables breeders to identify and rescue rare genetic variants with improved tolerance. Moreover, screening for desirable traits *in vitro* is much faster and requires less space than traditional field-based methods, accelerating the development of traits like disease or salt resistance [12]. Additionally, somaclonal variation observed during cucumber tissue culture, once seen as a mere drawback, is now intentionally utilized under controlled selection pressure to produce novel variation for crop improvement [10, 12, 13].

Haploid and doubled haploid (DH) plant production has emerged as a powerful tool in modern cucumber breeding programs, particularly in overcoming the limitations of traditional selection methods. Through tissue culture-based approaches such as anther culture (androgenesis) and irradiated pollen-induced gynogenesis, breeders can rapidly obtain completely homozygous lines within a single generation, significantly accelerating breeding cycles [14, 15]. Additionally, protoplast culture in *Cucumis sativus* represents a versatile *in vitro* technique that enables cell wall removal and subsequent regeneration into whole plants. Protoplast systems have been used for haploid and mixoploid isolation, as well as PEG-mediated somatic hybridization involving cucumber and related *Cucumis* species, enabling novel genetic combinations not achievable through sexual crossing [16].

Due to the limitations of traditional breeding and recent advances in biotechnology, *in vitro* tissue culture techniques have become essential in modern cucumber improvement. Main regeneration methods—somatic embryogenesis, organogenesis, and haploid plant production—allow for genetic fidelity, efficient shoot regeneration, and quick achievement of homozygosity, respectively [8, 9, 14]. Micropropagation supports large-scale multiplication of elite, disease-free lines [7, 17], while protoplast culture shows promise for genetic manipulation and somatic hybridization [18]. This chapter aims to offer a comprehensive overview of the key *in vitro* tissue culture techniques used to improve cucumber.

2. *In vitro* systems

2.1 Somatic embryogenesis

Somatic embryogenesis (SE) is a pivotal regeneration pathway in cucumber, enabling efficient clonal propagation, facilitating genetic transformation, and contributing to the development of stress-tolerant cultivars (**Figure 1**). The success of SE is strongly influenced by factors such as genotype, explant type, and *in vitro* culture conditions. Genotypic effects play a crucial role as limiting factors. Hamza, Diab [19] found that the Egyptian cultivar Beit Alpha could produce up to 60% of embryogenic callus from cotyledon-leaf explants, whereas the cultivar Prince only reached about 40%, emphasizing the importance of genotype in regeneration success. Similar differences in SE potential among various cultivars have been consistently documented in the existing literature [20, 21]. Cotyledons are the most frequently used and responsive explants for SE induction, providing high embryogenic potential when cultured on media enriched with Ref. [8, 22]. Hypocotyls and leaf segments also demonstrate potential, though they typically need more precise hormone balances to induce embryogenic callus [23, 24]. Additionally, shoot tips, mature seeds,

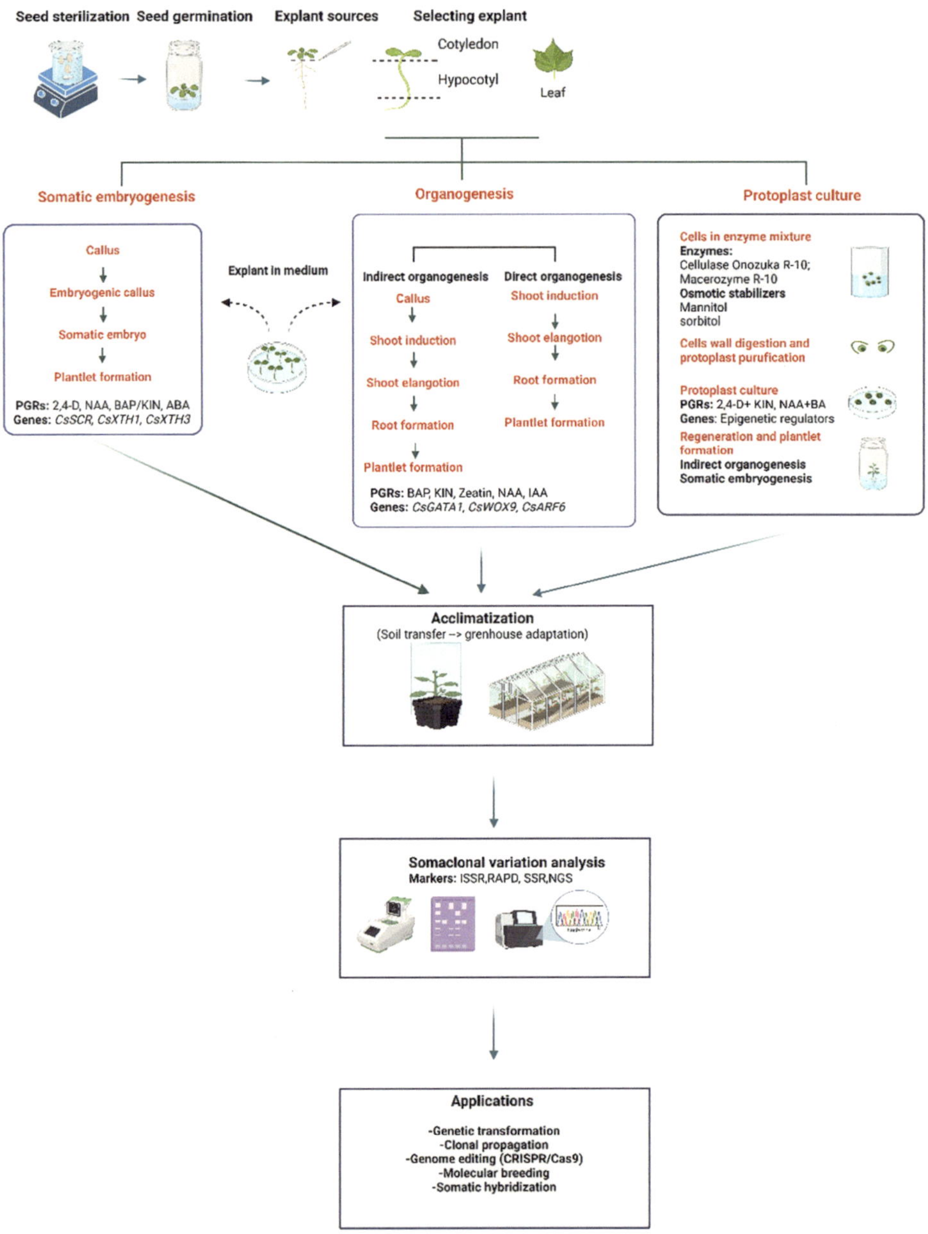

Figure 1.
*In vitro regeneration pathways in cucumber (*Cucumis sativus *L.) through somatic embryogenesis, organogenesis, and protoplast culture, followed by acclimatization, somaclonal variation analysis, and applications. Created with BioRender.com.*

and cotyledon-node regions have been studied, showing varied responses influenced mainly by the cucumber cultivar and the explant's developmental stage [25, 26].

Plant growth regulators (PGRs) are crucial for controlling somatic embryo initiation, development, and maturation. 2,4-Dichlorophenoxyacetic acid (2,4-D) remains the most common auxin used for SE induction in cucumber. Studies show that 2–5 mg/L of 2,4-D significantly encourages callus formation and early embryo development, especially in cotyledon and leaf explants [24, 27]. Naphthaleneacetic acid (NAA) is also used, often in combination with cytokinins such as benzylaminopurine (BAP) or kinetin (KIN), to enhance callus proliferation and facilitate the transition to somatic embryos [20, 26]. The latter stages of embryo maturation and shoot conversion typically require a hormonal shift toward cytokinin dominance. The use of abscisic acid (ABA) has also been shown to promote maturation and desiccation-like responses, enhancing the quality of somatic embryos [28]. Additionally, carbohydrate sources with high sucrose levels (up to 12%) not only supply energy but also create osmotic stress that mimics zygotic conditions, further stimulating embryogenesis [22, 29]. Additionally, bioreactors like RITA® have facilitated scaled-up SE systems in elite hybrids [30].

At the molecular level, several genes (e.g., *CsSCR*, *CsXTH1*, *CsXTH3*) and proteins have been identified as being upregulated during somatic embryo induction [21, 31]. Additionally, studies using inter-simple sequence repeats (ISSR) markers, RT-qPCR, and hormone profiling have clarified the transcriptional and hormonal pathways active during dedifferentiation and redifferentiation [28, 31, 32]. CsSCR (SCARECROW) is a GRAS-family transcription factor strongly upregulated in globular- and heart-stage somatic embryos and in rooting tissues. Wisniewska and Grabowska [31] showed elevated CsSCR transcript levels, confirmed via in situ hybridization and expression assays, implicating its role in radial tissue organization during embryo formation. Malinowski and Filipecki [33] conducted a comprehensive study on two xyloglucan endotransglucosylase/hydrolase (XTH) genes, *Cs-XTH1* and *Cs-XTH3,* in *Cucumis sativus*. These enzymes remodel cell walls by cutting and rejoining xyloglucan chains, a critical process during embryo formation. Moreover, both genes showed elevated transcript levels in somatic embryos compared to pre-embryogenic tissue. Northern blot and *in situ* hybridization revealed Cs XTH1 localized in presumptive cotyledon primordia, while Cs XTH3 was also active during embryonic development.

2.2 Organogenesis

Organogenesis is a fundamental *in vitro* regeneration strategy in cucumber, employed widely for clonal propagation and genetic transformation (**Figure 1**). The efficiency of organogenesis is influenced by several interacting factors, including the type of explant, culture medium, PGRs, genotype, and environmental conditions [4]. Genotypic variation plays a critical role in the success of organogenesis in cucumber. Different cultivars exhibit significant differences in their regenerative responses under identical *in vitro* conditions. Some genotypes exhibit high shoot regeneration frequencies from specific explants, such as cotyledons or shoot tips, while others respond poorly, even with optimized media and PGR. This variation is likely due to underlying genetic control of hormonal sensitivity, developmental plasticity, and expression of key regulatory genes. Wang and Zhou [34] identified candidate genes (e.g., CsGATA1) and quantified trait loci (QTLs) associated with regeneration ability, underscoring the importance of genotype selection and molecular screening in developing efficient organogenic protocols for cucumber improvement. Nie and Shan [35] demonstrated that floral organogenesis is regulated via the CsDFB1–CsPHB module, which influences local auxin distribution. Slomnicka and Cieplak [21] identified QTLs such as sr6.1A and sr6.1B for shoot regeneration from leaf explants and highlighted CsWOX9 and CsARF6 as candidate genes for organogenic control.

Among explants, cotyledonary nodes and cotyledons are the most responsive. Selvaraj and Vasudevan [36] achieved up to 86% of shoot regeneration from cotyledon explants using MS medium with BAP, NAA, zeatin, and glutamine. Liu [37] demonstrated that cotyledon nodes yielded a higher adventitious bud induction rate compared to true leaves, radicles, and hypocotyls. Zheng [38] reported that changing the explant source from stem nodes to cotyledons improved the bud induction rate from 63.5% to 78.3%. Similarly, Venkatachalam and Jinu [7] demonstrated over 50 shoots per explant using 1.5 mg/L BAP + 4.5 mg/L silver nitrate in cotyledonary nodes, confirming their high morphogenic capacity. Hypocotyls and shoot tips are also used. Sangeetha and Venkatachalam [39] reported direct shoot formation and *in vitro* flowering from shoot tips on BAP-enriched medium. Although explants such as hypocotyls and leaves can also support regeneration, they generally exhibit lower efficiency than cotyledon-derived tissues [4]. Furthermore, the explant's physiological age is crucial in determining its regenerative ability. Young seedlings, typically 1 to 7 days old, are generally more suitable for organogenesis because their cells are less differentiated and respond more effectively to *in vitro* signals. Explants at these early stages can improve shoot induction rates and overall transformation success, owing to increased cellular plasticity and hormonal responsiveness [40].

PGRs play a crucial role in initiating and directing the organogenic process. Cytokinnins such as BAP, KIN, and zeatin stimulate shoot bud formation, while auxins like NAA and indole-3-acetic acid (IAA) help in callus induction and shoot elongation [41]. For hypocotyls, optimal shoot regeneration was achieved using 3.6 μM 2,4-D and 2.2 μM BAP [36]. ABA has been reported to prevent vitrification and improve organogenesis quality [4]. Moreover, silver nitrate enhances shoot quality by mitigating the effects of ethylene [42]. Venkatachalam and Jinu [7] showed that a medium containing 1.5 mg/L BAP and 4.5 mg/L silver nitrate significantly improved shoot proliferation from cotyledonary node explants, yielding over 50 shoots per explant and reducing hyperhydricity. In addition, environmental factors also play critical roles in plant regeneration. Temperature, gelling agent type, and photoperiod length have been shown to influence morphogenesis outcomes [43]. Light-induced transcriptional responses and NO-mediated cell cycle activation also contribute to shoot organogenesis in cucumber [44, 45].

2.3 Haploid culture

Haploid and DH production techniques have revolutionized plant breeding by enabling the rapid fixation of homozygosity within a single generation, significantly accelerating the development of pure lines. In cucumber, a cross-pollinated and highly heterozygous species, generating haploid and DH plants, offers substantial advantages for hybrid seed production, QTL mapping, mutation screening, and functional genomics. Although these methods are well established in cereals and solanaceous crops, their efficient implementation in cucurbits has been more challenging due to species-specific resistance, low regeneration rates, and strong genotype dependence [46, 47]. Nevertheless, two decades of intensive research have led to significant advances in haploid induction, utilizing anther culture, ovary culture, and parthenogenesis, supported by improvements in explant selection, media formulation, pretreatment protocols, and molecular techniques validation (**Figure 2**).

Among these approaches, anther culture remains one of the most studied methods for inducing haploid formation via androgenesis. Microspores at the uninucleate to early binucleate stages, typically found in 2–4 mm long floral buds, are considered

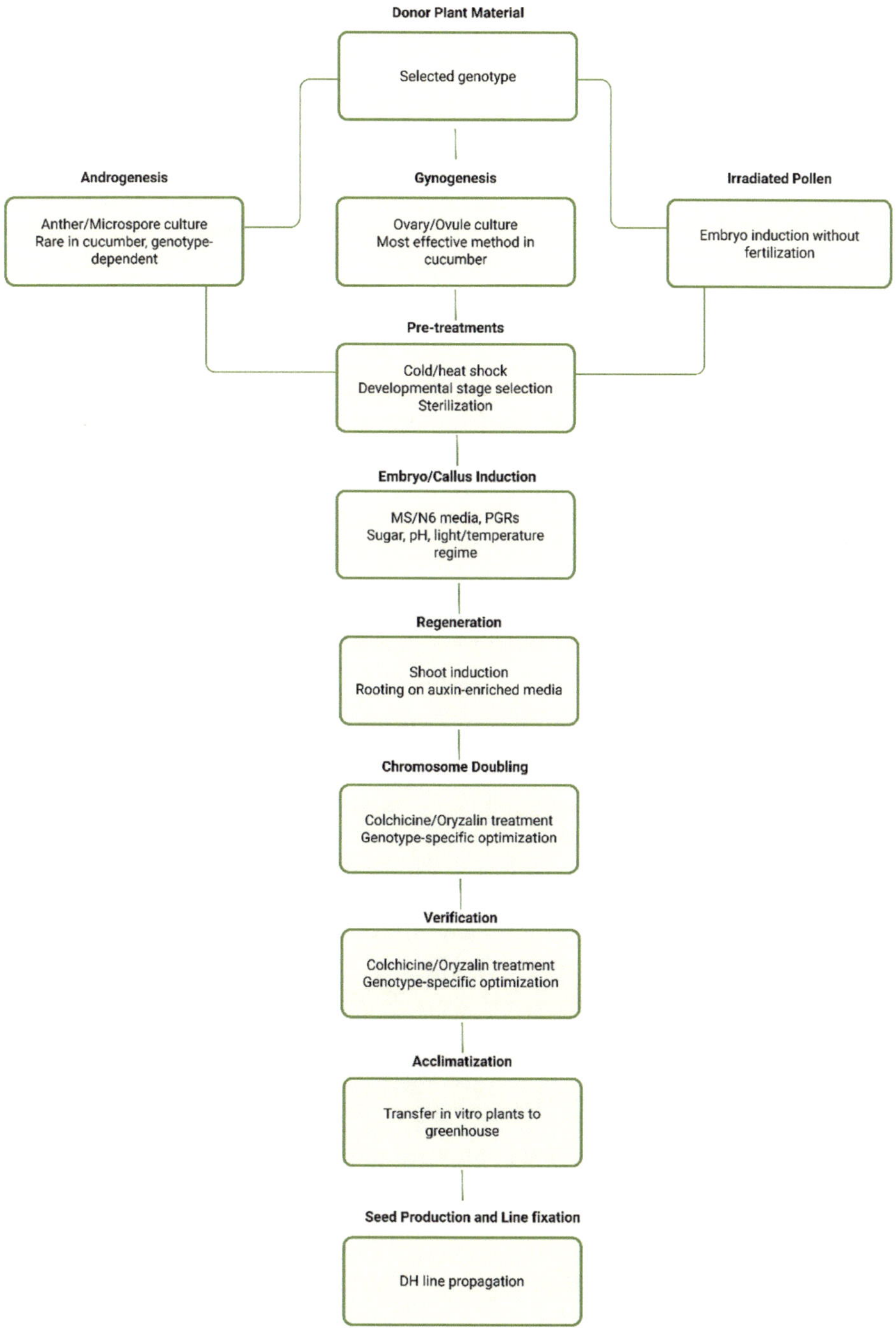

Figure 2.
*Doubled haploid (DH) production in cucumber (*Cucumis sativus *L.). Created with BioRender.com.*

most responsive to *in vitro* embryogenic induction [14, 48]. However, cucumber is a recalcitrant species for androgenesis due to the frequent formation of non-embryogenic callus and albino plantlets [49, 50]. These limitations have stimulated the exploration of alternative explants and pathways. Gynogenesis, through the culture of unpollinated ovaries or ovules, has emerged as a viable and sometimes superior alternative, especially for genotypes with poor androgenic response. Ovary explants collected 1 to 3 days after anthesis have shown high embryogenic potential under appropriate culture conditions [51, 52]. Domblides and Shmykova [53] further demonstrated the success of unpollinated ovule culture in various cucurbit species, reinforcing the relevance of gynogenesis for haploid production. Furthermore, *in vitro* ovary culture can circumvent the developmental limitations of male gametophyte-derived explants, especially in genotypes that are poor responders to another culture [51, 54]. In addition to androgenesis and gynogenesis, parthenogenesis, embryo development from unfertilized egg cells, represents another viable strategy for haploid production. Parthenogenetic haploid embryos can be induced either *in vivo*, using irradiated pollen or distant hybridization (e.g., melon × cucumber), or *in vitro* through hormone treatments. Parkash and Bhatia [55] reported an efficient protocol for *in vitro* induction of parthenogenetic embryos using hormone-enriched MS media. Similarly, Salehian and Shahnazi [56] achieved DH plant production through parthenogenesis combined with precise embryo rescue techniques. These approaches are particularly advantageous in breeding programs aiming to develop DH lines for hybrid seed production. In particular, the application of cycocel (CCC) in combination with polyamines in the culture medium has been effective in stimulating parthenogenesis and improving the regeneration of haploid plantlets [57].

The choice of culture medium and hormonal balance plays a central role in determining the success of haploid induction. Murashige and Skoog (MS) medium remains the most commonly used basal medium, although modifications such as reducing nitrate concentration or adding vitamins (e.g., thiamine, myo-inositol) have been tested to optimize callogenesis and embryogenesis [51]. PGRs are critical for directing the developmental pathway of explants. Cytokinins such as BAP, KIN, and especially thidiazuron (TDZ) have been widely used to enhance shoot regeneration and embryogenic callus formation. The use of MS medium supplemented with silver nitrate, TDZ, and KIN has been shown to promote gynogenic embryo induction [58]. On the auxin side, NAA and 2,4-D are often employed during the callus induction phase to support cell dedifferentiation and proliferation [51]. Moreover, the incorporation of organic additives such as casein hydrolysate, coconut water, or activated charcoal has shown potential in improving the quality and frequency of embryogenic cultures [57]. These components help stabilize the pH, absorb phenolic compounds, and provide additional nutrients or growth-promoting substances.

In vitro embryogenesis is often facilitated by the application of stress treatments that redirect gametophytic development toward sporophytic pathways. Cold pretreatment, involving exposure of flower buds or ovary explants to low temperatures (typically 4°C) for 2–7 days, has consistently been shown to enhance embryogenic potential in cucumber [59]. The mechanism is believed to involve stress-induced expression of embryogenesis-related genes and hormonal reprogramming. Additionally, osmotic stress through the use of high sucrose or mannitol concentrations, or heat shock treatments, can improve embryogenic outcomes by inducing physiological changes in microspores or ovular tissues. In androgenesis protocols, a short heat shock at 32–35°C is sometimes applied prior to culture initiation [60].

For the practical application of haploid plants in breeding, chromosome doubling is required to restore fertility. This is typically achieved through the application of antimitotic agents such as colchicine, oryzalin, or trifluralin, which disrupt spindle formation during mitosis. Ebrahimzadeh and Shariatpanahi [57] demonstrated successful chromosome doubling in cucumber haploids by soaking regenerated shoots in colchicine or trifluralin solutions, achieving a high percentage of fertile, morphologically normal DH plants. Shoot tip immersion and *in vitro* treatments are commonly used to avoid toxicity and ensure uniform exposure. The timing and duration of exposure must be carefully optimized to balance efficacy with plant survival. Also, verification of haploid and DH status is a crucial step in haploid breeding. Flow cytometry is a rapid and reliable method for ploidy analysis, allowing high-throughput screening of regenerated lines. Molecular markers such as simple sequence repeats (SSRs), random amplified polymorphic DNA (RAPD), and ISSRs are also employed to assess homozygosity and confirm the genetic uniformity of DH lines [47, 61].

2.4 Protoplast culture

Protoplast culture in cucumber has emerged as a powerful biotechnological tool for genetic transformation, somatic hybridization, and *in vitro* regeneration, despite being limited in routine applications due to technical complexities and genotype-dependent variability (**Figure 1**) [4, 62]. Protoplasts, plant cells devoid of their cell walls, serve as an excellent model system for studying cellular totipotency, early morphogenesis, and cell wall regeneration under controlled *in vitro* conditions [18, 63]. Studies have demonstrated successful plant regeneration via direct somatic embryogenesis and organogenesis from mesophyll and cotyledon-derived protoplasts in cucumber. However, the frequency and stability of regeneration are often highly genotype-specific [64, 65].

Efficient isolation and culture of cucumber protoplasts have been reported using the enzymatic digestion of cotyledon or mesophyll tissues. Huang and Wang [18] developed a high-yield protoplast isolation method using a combination of 1.5% cellulase R10 and 0.4% macerozyme R10, supplemented with 0.6 M mannitol as an osmotic stabilizer in an enzyme solution buffered with MES hydrate (2-(N-morpholino hydrate) ethane sulfonic acid at pH 5.7. The digestion was conducted under dark conditions with gentle shaking at 25°C for 6 hours, resulting in highly viable protoplasts suitable for transient gene expression and downstream applications. Similarly, Punja and Tang [62] achieved successful isolation and regeneration of protoplasts derived from both mesophyll and cotyledon tissues using a two-enzyme mixture of cellulase and pectinase in a modified KM (kao and michayluk) medium with 0.4–0.5 M sorbitol. Their results demonstrated that shoot regeneration was possible through plating protoplasts onto solidified KM medium enriched with appropriate PGRs, such as 1 mg/L NAA and 0.5 mg/L BA. Furthermore, Burza and Malepszy [64] reported successful regeneration via direct somatic embryogenesis from cucumber protoplasts isolated using 1% cellulase Onozuka R-10 and 0.5% pectinase, suspended in an osmoticum containing 0.5 M glucose. They highlighted that embryogenic calli could be induced on N6-based medium supplemented with 2,4-D (2.0 mg/L) and KIN (0.5 mg/L), followed by shoot regeneration upon transfer to a cytokinin-enriched MS medium. These studies collectively emphasize the critical role of enzyme composition, osmotic balance, and hormone supplementation in establishing effective protoplast culture and regeneration systems in cucumber.

Protoplast regeneration is highly genotype-dependent, and the ability to form callus or regenerate shoots varies significantly among cucumber cultivars. Gajdová and Navrátilová [66] emphasized the importance of donor tissue type, culture medium composition, and genetic background in determining regeneration efficiency across both *Cucumis* and *Cucurbita* species. Their study demonstrated that cotyledon and mesophyll tissues often yield higher protoplast viability and regenerative capacity compared to other tissues. Moreover, they showed that some genotypes possess a strong embryogenic potential, whereas others exhibit limited or no callus formation under identical conditions. Jia and Fu [67] were among the first to successfully achieve plant regeneration through somatic embryogenesis from protoplasts isolated from cotyledon tissue in cucumber. Their protocol involved a culture medium enriched with auxins (such as 2,4-D) and cytokinins (like KIN), which promoted initial cell wall formation, subsequent callus induction, and finally embryoid development. Moreover, reactive oxygen and nitrogen species also play critical roles in protoplast viability and regeneration capacity. Petrivalsky and Vanícková [68] demonstrated that appropriate management of oxidative stress during early protoplast culture can significantly enhance regeneration outcomes. This pioneering work established a foundational platform for further advancements in cucumber protoplast regeneration systems.

At the molecular level, Cápal and Ondřej [63] investigated the gene expression profiles and chromatin changes in cucumber protoplast cultures. They reported that dedifferentiation and cell cycle re-entry were associated with alterations in chromatin structure, particularly histone modifications and chromatin relaxation. These changes facilitate the transition of isolated protoplasts back to a pluripotent state, which is essential for initiating cell division and morphogenesis. Supporting this, Moricova and Ondrej [69] demonstrated that protoplast culture leads to significant changes in DNA methylation (5-mC) and hydroxymethylation (5-hmC) levels. Their findings suggest that epigenetic reprogramming plays a crucial role in successful regeneration, as these modifications are likely involved in reactivating genes required for cell proliferation and differentiation. High levels of 5-hmC, in particular, were linked to the early stages of organogenesis and could serve as epigenetic markers of regenerative potential.

Protoplast fusion has emerged as an effective method to overcome sexual incompatibility barriers in cucumber breeding by enabling somatic hybridization between *Cucumis* species. In their pioneering work, Fellner and Lebeda [70] successfully isolated and fused protoplasts derived from *Cucumis sativus* and *Cucumis melo*, two species that are sexually incompatible via conventional crossbreeding. Protoplasts were enzymatically isolated from leaf mesophyll tissues and fused using polyethylene glycol (PEG)-mediated fusion, a common technique that induces membrane fusion and cytoplasmic mixing. The fused cells were cultured on optimized media to regenerate hybrid calli. Although complete plant regeneration was not reported in this initial study, the formation of heterokaryons and hybrid cell colonies demonstrated that interspecific gene transfer was feasible, offering potential for the introgression of melon traits (e.g., disease resistance or fruit quality) into cucumber. Skálová and Navrátilová [16] explored haploid and mixoploid cucumber protoplasts, focusing on their isolation, fusion behavior, and chromosomal constitution. The researchers used a highly purified enzymatic mixture (typically cellulase and macerozyme) to isolate protoplasts from in vitro-grown cucumber tissues, followed by fluorescent staining and flow cytometry to determine ploidy levels post-fusion. PEG-mediated fusion was again used to generate somatic hybrids, and the resulting mixoploid protoplasts were assessed for cell division and potential to regenerate. Their analysis showed successful fusion and partial chromosome doubling, suggesting that this method could be used to develop polyploid lines or somatic hybrids with novel trait combinations. Importantly, they demonstrated that ploidy manipulation via fusion can be monitored and stabilized during early culture phases critical for downstream regeneration and breeding applications.

3. Somaclonal variation in cucumber

Somaclonal variation refers to the genetic or epigenetic changes that occur during *in vitro* plant regeneration, often from callus or suspension cultures. It can manifest as chromosomal mutations, gene sequence alterations, or methylation changes [71]. In cucumber, this phenomenon presents both a challenge and an opportunity, posing potential risks for clonal fidelity during micropropagation while also offering a valuable source of new genetic diversity for breeding. These variations, which can be genetic or epigenetic, often originate from callus or suspension cultures and may appear as chromosomal abnormalities, gene mutations, transposon activation, or DNA methylation changes. Several studies have demonstrated that explant type, physiological status, culture conditions and duration, and especially medium composition including PGRs significantly influence the level and type of somaclonal changes [12, 72]. To exploit this variation effectively, precise culture protocols are essential. This includes minimizing the number of subculture cycles since prolonged *in vitro* exposure correlates with higher variation rates avoiding excessive callus proliferation phases and carefully regulating the concentrations of PGRs [73].

In cucumber, somatic embryogenesis and callus-based regeneration are widely used in tissue culture protocols, and these methods have been linked with considerable variation among regenerants. Custers and Zijlstra [74] observed that although R_1 somaclones displayed morphological variability, most aberrations were transient and eliminated through sexual reproduction, with only a few stable variants persisting into the R_2 generation. However, modern molecular tools such as next-generation sequencing have revealed extensive underlying genomic variation in cucumber somaclones including thousands of single nucleotide polymorphisms (SNPs) and structural variants despite relatively subtle phenotypic manifestations [10, 75].

Pawelkowicz and Skarzynska [10] provided deep transcriptomic insights by comparing three somaclonal cucumber lines (S1–S3), derived from the same inbred line (Borszczagowski B10) but regenerated via different *in vitro* pathways direct organogenesis, callus-mediated regeneration, and embryogenic suspension culture. Each somaclonal line displayed distinct morphological and transcriptional profiles. Hundreds of differentially expressed genes (DEGs) were identified, related to stress response, metabolic processes, DNA/RNA binding, and transcriptional regulation. Notably, the DEGs overlapped partially across lines, indicating the existence of conserved molecular targets responsive to culture-induced stress. Moreover, several DEGs were associated with SNPs, supporting the role of both genetic and epigenetic mechanisms in driving somaclonal variation. Earlier studies corroborate these findings. Plader and Malepszy [76] and Ladyzynski and Burza [77] demonstrated that cytokinin-rich media, especially those containing BAP, increased the frequency of chromosomal abnormalities and ploidy shifts in cucumber regenerants. Bartoszewski and Havey [78] further reported alterations in organelle genomes in somaclonal lines, underscoring the broad impact of culture conditions on genomic integrity.

While somaclonal variation introduces uncertainty in tissue culture applications where genetic fidelity is essential (e.g., micropropagation or germplasm conservation), it can be deliberately harnessed as a tool for genetic improvement in cucumber breeding. Controlled induction of somaclonal variation has been shown to generate novel traits such as altered fruit shape and size, enhanced tolerance to abiotic stressors (e.g., salinity, drought), and increased disease resistance—all without the need for transgenic modification [12]. These variants serve as a valuable resource in breeding programs, particularly for hybrid development and functional genomics studies.

Unlike conventional mutation breeding or genetic engineering, somaclonal variation offers a non-genetically modified organism (non-GMO) alternative that may be more readily accepted in regions with regulatory constraints or consumer resistance.

4. Transgenic approaches and CRISPR/Cas9 genome editing in cucumber

Cucumber is an essential member of the Cucurbitaceae family, serving as a model for studying sex determination, phloem biology, and fruit development. Despite the availability of high-quality genomic resources, conventional breeding has limited success in developing resistant cultivars due to limited natural variation and sexual incompatibility barriers. As a result, transgenic and genome-editing strategies have gained momentum for introducing or modifying desirable traits. CRISPR/Cas9 technology in particular allows for efficient and targeted gene editing, representing a breakthrough in cucumber biotechnology.

Genetic transformation in cucumber has primarily relied on *Agrobacterium tumefaciens*-mediated protocols. Cotyledonary explants are the most frequently used starting material due to their high regenerative potential. Zhao and Qi [79] demonstrated an optimized transformation system using proximal cotyledon explants of 2-day-old seedlings. Their protocol included pre-culturing on MS medium supplemented with 1.0 mg/L BAP and 0.5 mg/L ABA, followed by infection with Agrobacterium strain EHA105 in the presence of 100 μM acetosyringone. The selection and regeneration media contained kanamycin and meropenem, leading to a transformation efficiency of 4.67%. Xin and Tian [80] further improved this process by integrating a vacuum infiltration step, significantly enhancing transformation efficiency across different cucumber cultivars. They also showed that transformation success was strongly genotype-dependent, suggesting the need for tailored protocols for elite lines. Nanasato and Konagaya [81] also developed an effective system using the EHA105 strain and cotyledon explants precultured in a BAP-rich medium. Their regeneration system supported the recovery of transgenic lines at frequencies exceeding 8%. In addition, Nishibayashi and Kaneko [82] used hypocotyl explants for transformation and noted comparable results to cotyledon explants, suggesting broader options in explant source selection. In addition to BAP and ABA, various combinations of cytokinins (TDZ, KIN) and auxins (NAA, IAA) have been tested to optimize shoot induction and rooting in transgenic tissues. Ni and Li [83] successfully used such protocols to integrate the nattokinase gene into cucumber plants using CRISPR/Cas9.

CRISPR/Cas9 technology has enabled precise gene knockouts in cucumber, particularly for disease resistance. The mildew resistance locus O (MLO) gene family has been a major focus due to its negative regulatory role in powdery mildew resistance. Tek and Calis [84] and Shnaider and Elad [85] reported successful editing of CsaMLO8, which conferred strong resistance to Podosphaera xanthii without compromising plant morphology. In the viral disease context, Fidan and Calis [86] disrupted the *eIF4E* gene using CRISPR/Cas9. This gene is a known susceptibility factor for several potyviruses. Edited cucumber lines displayed resistance to WMV, ZYMV, and PRSV, and importantly, these plants maintained high fertility and morphological integrity. Chandrasekaran and Brumin [87] laid the groundwork for virus resistance via targeted genome editing by demonstrating that eIF4E mutants are broadly resistant to RNA viruses. Their work influenced further studies targeting translation initiation factors in cucurbits. Zheng and Yang [88] targeted CsNBS-LRR gene families involved in bacterial wilt resistance and reported successful gene editing with significant reductions in disease symptoms.

Beyond disease resistance, genome editing technologies have been increasingly applied in cucumber to engineer metabolic traits, enabling nutritional enhancement and the production of high-value bioactive compounds. Ni and Li [83] introduced the *nattokinase* gene into cucumber plants using CRISPR/Cas9. This not only demonstrated the crop's amenability to foreign gene insertion but also highlighted cucumber's potential as a molecular farming platform for nutraceutical and pharmaceutical production. Additionally, Mitra and Shnaider [89] used CRISPR/Cas9 to study genes involved in fruit set and development. Knockouts of candidate genes like *CsGA3ox* and *CsIAA9* revealed their regulatory roles, allowing manipulation of fruit formation pathways. Moreover, recent studies have expanded CRISPR's application to abiotic stress tolerance. Peng and Chen [90] identified *CsAKT1* as a key gene involved in salt tolerance, which was validated using CRISPR/Cas9-edited lines. The knockout mutants displayed altered potassium uptake and enhanced salt resistance. Dong and Li [91] focused on the *STAYGREEN* gene, showing its role in low-temperature tolerance. Genome-edited lines exhibited improved chlorophyll retention and better cold tolerance.

To accelerate gene function studies, Nguyen and Hoang [92] developed a hairy root transformation system using *Agrobacterium rhizogenes*. This method allows the rapid assessment of CRISPR constructs without the lengthy regeneration process. Transgenic hairy roots expressing Cas9-sgRNA cassettes successfully induced mutations in target genes within 10–14 days. Kiryushkin and Ilina [93] demonstrated hairy root-based CRISPR editing for root architecture studies. Their work provides a foundation for applying root-focused gene manipulation, particularly under stress or nutrient-deficient conditions.

5. Conclusion

In vitro culture technologies combined with modern genetic tools have transformed cucumber improvement. Using regeneration systems like organogenesis, somatic embryogenesis, haploid induction, and micropropagation, researchers have built platforms for trait introgression, rapid propagation, and genome editing. Explants such as cotyledons and hypocotyls, with optimized hormone combinations (e.g., BAP, NAA, TDZ), enable high-efficiency shoot and root regeneration, which is foundational for transformation. Advances in Agrobacterium-mediated transformation ensure stable gene integration, while CRISPR/Cas9 offers precise editing for functional genomics. Editing genes related to biotic (e.g., *MLO*, *eIF4E*) and abiotic stress (e.g., *CsAKT1*, *CsSGR*) yields lines with better resistance and adaptability. Metabolic engineering for traits such as β-carotene and vitamin C highlights cucumber's potential for functional foods. Challenges include genotype dependence, limited protocols for elite cultivars, and unclear regulatory guidelines. Future focus should be on genotype-independent systems, improved CRISPR delivery, and omics integration for multi-target editing. Overall, combining *in vitro* methods with genome editing promises to enhance cucumber breeding, thereby increasing yield, resilience, and nutritional value for global agriculture.

Author details

Sumeyra Ucar[1], Esma Yigider[2], Esra Yaprak[1], Ayse Gul Kasapoglu[1], Melek Ekinci[3], Ertan Yildirim[3] and Murat Aydin[2]*

1 Department of Molecular Biology and Genetics, Erzurum Technical University, Erzurum, Turkey

2 Faculty of Agriculture, Department of Agricultural Biotechnology, Atatürk University, Erzurum, Turkey

3 Faculty of Agriculture, Department of Horticulture, Atatürk University, Erzurum, Turkey

*Address all correspondence to: maydin@atauni.edu.tr

References

[1] Kaur M, Sharma P. Recent advances in cucumber (*Cucumis sativus* L.). The Journal of Horticultural Science and Biotechnology. 2021;**97**(1):3-23

[2] Javid H, Fatima U, Rukhsar A, Hussain S, Bibi S, Bodlah MA, et al. Phytochemical, nutritional and medicinal profile of *Cucumis sativus* L. (cucumber). Food Science and Engineering. 2024;**5**(2):358-377

[3] Anwar S, Siddique R, Ahmad S, Haider MZ, Ali H, Sami A, et al. Genome wide identification and characterization of Bax inhibitor-1 gene family in cucumber (*Cucumis sativus*) under biotic and abiotic stress. BMC Genomics. 2024;**25**(1):1032

[4] Tan J, Lin L, Luo H, Zhou S, Zhu Y, Wang X, et al. Recent Progress in the Regeneration and Genetic Transformation System of Cucumber. Applied Sciences. 2022;**12**:7180. DOI: 10.3390/app12147180

[5] Naegele RP, Wehner TC. Genetic resources of cucumber. Genetics and Genomics of Cucurbitaceae. 2017:61-86

[6] Mirzwa-Mroz E, Zieniuk B, Yin Z, Pawelkowicz M. Genetic insights and molecular breeding approaches for downy mildew resistance in cucumber (*Cucumis sativus* L.): Current Progress and future prospects. International Journal of Molecular Sciences. 2024;**25**(23):12726

[7] Venkatachalam P, Jinu U, Sangeetha P, Geetha N, Sahi SV. High frequency plant regeneration from cotyledonary node explants of *Cucumis sativus* L. cultivar 'green long' via adventitious shoot organogenesis and assessment of genetic fidelity by RAPD-PCR technology. 3 Biotech. 2018;**8**(1):60

[8] Dhumal SS, Naik BV, Nimbalkar MS. Advances in tissue culture of cucurbits: A review. International Journal of Current Microbiology and Applied Sciences. 2020;**9**:2887-2910

[9] Nanasato Y, Tabei Y. A method of transformation and current progress in transgenic research on cucumbers and Cucurbita species. Plant Biotechnology (Tokyo). 2020;**37**(2):141-146

[10] Pawelkowicz ME, Skarzynska A, Mróz T, Bystrzycki E, Plader W. Molecular insight into somaclonal variation phenomena from transcriptome profiling of cucumber (*Cucumis sativus* L.). Plant Cell, Tissue and Organ Culture. 2021;**145**(2):239-259

[11] Dorani E, Dehghanian Z, Gougerdchi V, Hamedpour-Darabi M. Application of Somaclonal variation in crop improvements. In: Plant Mutagenesis: Sustainable Agriculture and Rural. Landscapes: Springer; 2024. pp. 93-109

[12] Krishna H, Alizadeh M, Singh D, Singh U, Chauhan N, Eftekhari M, et al. Somaclonal variations and their applications in horticultural crops improvement. 3 Biotech. 2016;**6**(1):54

[13] Ferreira MD, Rocha AD, Nascimento FD, Oliveira WDD, Soares JMD, Rebouças TA, et al. The role of Somaclonal variation in plant genetic improvement: A systematic review. Agronomy-Basel. 2023;**13**(3):730

[14] Asadi A, Segui-Simarro JM. Production of doubled haploid plants in cucumber (*Cucumis sativus* L.) through

anther culture. Methods in Molecular Biology. 2021;**2289**:71-85

[15] Pradeepkumara N, Parkash C, Bhatia R, Munshi AD, Rao M, et al. An efficient and easy-to-use protocol for induction of haploids in cucumber through parthenogenic embryo development. Botanical Studies. 2024;**65**(1):30

[16] Skálová D, Navrátilová B, Dolezalová I, Vasut RJ, Lebeda A. Haploid and mixoploid cucumber (*Cucumis sativus* L.) protoplasts - isolation and fusion. Journal of Applied Botany and Food Quality. 2012;**85**(1):64-72

[17] Ahmad N, Anis M. In vitro mass propagation of *Cucumis sativus* L. from nodal segments. Turkish Journal of Botany. 2005;**29**(3):237-240

[18] Huang HY, Wang ZY, Cheng JT, Zhao WC, Li X, Wang HY, et al. An efficient cucumber (*Cucumis sativus* L.) protoplast isolation and transient expression system. Scientia Horticulturae Amsterdam. 2013;**150**:206-212

[19] Hamza HM, Diab RH, Khatab IA, Gaafar RM, Elhiti M. Enhancing in vitro regeneration via somatic embryogenesis and Fusarium wilt resistance of Egyptian cucumber (*Cucumis sativus* L.) cultivars. Journal, Genetic Engineering & Biotechnology. 2024;**22**(1):100360

[20] Jesmin R, Mian MAK. Callus induction and efficient plant regeneration in cucumber (*Cucumis sativus* L.). Journal of Bioscience and Agriculture Research. 2016;**9**(02):796-803

[21] Slomnicka R, Cieplak M, Antosiewicz M, Sadlos A, Galczak A, Kazminska K, et al. Identification of quantitative trait loci for in vitro plant regeneration from leaf microexplants in cucumber (*Cucumis sativus* L.). Journal of Applied Genetics. 2024;**66**(3):545-555

[22] Lou H, Kako S. Role of high sugar concentrations in inducing somatic embryogenesis from cucumber cotyledons. Scientia Horticulturae-Amsterdam. 1995;**64**(1-2):11-20

[23] Ugandhar T, Srilatha T, Imran M. Callus induction and somatic embryogenesis from leaf explants of cucumber (*Cucumis sativus* L.). International Journal of Integrative Sciences, Innovation and Technology. (IJIIT); 2013;**2**(2):29-33

[24] Usman M, Hussain Z, Fatima B. Somatic embryogenesis and shoot regeneration induced in cucumber leaves. Pakistan Journal of Botany. 2011;**43**(2):1283-1293

[25] Abu-Romman S, Suwwan M, Ezz A. The influence of plant growth regulators on callus induction from hypocotyls of cucumber (*Cucumis sativus* L.). Advances in Environmental Biology. 2013;**7**(2):339-344

[26] Khidr Y, El-Absawy E-S, Mahmoud A, Hasan M, Hemeida A. Somatic embryogenesis and plant regeneration induced from mature seeds, cotyledons and shoot tips of cucumber. Journal of Productivity and Development. 2012;**17**(2):193-209

[27] Mashayekhi K, Sharifani M, Shahsavand M, Kalati H. Induction of somatic embryogenesis in absence of exogenous auxin in cucumber (*Cucumis sativus* L.). International Journal of Plant Production. 2008;**2**(2):163-166

[28] Gawronska H, Burza W, Bolesta E, Malepszy S. Zygotic and somatic embryos of cucumber (*Cucumis sativus* L.) substantially differ in their

levels of abscisic acid. Plant Science. 2000;**157**(1):129-137

[29] Ladyman JAR, Girard B. Cucumber somatic embryo development on various gelling agents and carbohydrate sources. HortScience. 1992;**27**(2):164-165

[30] Kingidila M, Hunter A, Hennerty M. Production of F1 hybrid cucumber (*Cucumis sativus* L.) cultivars via somatic embryogenesis using a RITA® bioreactor. Acta Horticulturae. 2010;**923**:25-32. DOI: 10.17660/ActaHortic.2011.923.2

[31] Wisniewska A, Grabowska A, Pietraszewska-Bogiel A, Tagashira N, Zuzga S, Woycicki R, et al. Identification of genes up-regulated during somatic embryogenesis of cucumber. Plant Physiology and Biochemistry. 2012;**50**(1):54-64

[32] Xue W, Liu N, Zhang T, Li J, Chen P, Yang Y, et al. Substance metabolism, IAA and CTK signaling pathways regulating the origin of embryogenic callus during dedifferentiation and redifferentiation of cucumber cotyledon nodes. Scientia Horticulturae Amsterdam. 2022;**293**:110680

[33] Malinowski R, Filipecki M, Tagashira N, Wisniewska A, Gaj P, Plader W, et al. Xyloglucan endotransglucosylase/hydrolase genes in cucumber (*Cucumis sativus*) - differential expression during somatic embryogenesis. Physiologia Plantarum. 2004;**120**(4):678-685

[34] Wang Y, Zhou Q, Zhu G, Wang S, Ma Y, Miao H, et al. Genetic analysis and identification of a candidate gene associated with in vitro regeneration ability of cucumber. Theoretical and Applied Genetics. 2018;**131**(12):2663-2675

[35] Nie J, Shan N, Liu H, Yao X, Wang Z, Bai R, et al. Transcriptional control of local auxin distribution by the CsDFB1-CsPHB module regulates floral organogenesis in cucumber. Proceedings of the National Academy of Sciences of the United States of America. 2021;**118**(8):e2023942118

[36] Selvaraj N, Vasudevan A, Manickavasagam M, Ganapathi A. Organogenesis and plant formation in cucumber. Biologia Plantarum. 2006;**50**(1):123-126

[37] Liu Y. Study on Transfer of Csgrx4 Gene by Agrobacterium-Mediated into Cucumber [master's Thesis]. Yangling: Northwest A&F University; 2016

[38] Zheng L. Study on Regeneration and Agrobacterium-Mediated Genetic Transformation of Cucumber (*Cucumis sativus* L.). China: Yangzhou University Yangzhou; 2009

[39] Sangeetha P, Venkatachalam P. Induction of direct shoot organogenesis and flowering from shoot tip explants of cucumber (*Cucumis sativus* L. cv. 'Green long'). In vitro Cellular & Developmental Biology-Plant. 2014;**50**(2):242-248

[40] Chang H, Ni L, Sun Y, Luo W. Factors influencing agrobacterium-mediated genetic transformation of cucumber. Northern Horticulture. 2018;**2**:9-14

[41] Alam MF, Mondol PC, Roy K, Anisuzzaman M, Parvez MS, Ray SK, et al. Determination of apposite plant regeneration protocol for several cucurbits through direct and indirect organogenesis. The Horticultural Biotechnology Research. 2019;**5**:4-13

[42] Amirian R, Azadi P, Hojati Z. The effects of abscisic acid, silver nitrate and cefotaxime on cucumber in vitro organogenesis and evaluation of regenerated clones by AFLP marker.

Journal of Plant Process and Function. 2020;**9**(38):213-227

[43] Grozeva S, Velkov N. Plant regeneration of two cucumber (*Cucumis sativus* L.) genotypes: Effects of explant types and culture medium. Genetika-Belgrade. 2014;**46**(2):485-493

[44] Miguel JF. Effect of light conditions on in vitro adventitious organogenesis of cucumber cultivars. International Journal of Plant, Animal and Environmental Sciences. 2021;**12**(2022):138-144

[45] Zhu Y, Liao W, Niu L, Wang M, Ma Z. Nitric oxide is involved in hydrogen gas-induced cell cycle activation during adventitious root formation in cucumber. BMC Plant Biology. 2016;**16**(1):146

[46] Segui-Simarro JM, Moreno JB, Fernández MG, Mir R. Doubled haploid technology. Methods in Molecular Bsiology. 2021;**2287**:50

[47] Smiech M, Sztangret-Wisniewska J, Galecka T, Korzeniewska A, Marzec L, Kolakowska G, et al. Potential use of RAPD markers in characteristics of cucumber (*Cucumis sativus* L) haploids and double-haploids. Acta Societatis Botanicorum Poloniae. 2008;**77**(1):29-34

[48] Abdollahi MR, Najafi S, Sarikhani H, Moosavi SS. Induction and development of anther-derived gametic embryos in cucumber (*Cucumis sativus* L.) by optimizing the macronutrient and agar concentrations in culture medium. Turk. Journal of Biology. 2016;**40**(3):571-579

[49] Kumar HA, Murthy H, Paek K. Embryogenesis and plant regeneration from anther cultures of Cucumis sativus L. Scientia Horticulturae-Amsterdam. 2003;**98**(3):213-222

[50] Song H, Lou Q-F, Luo X-D, Wolukau JN, Diao W-P, Qian C-T, et al. Regeneration of doubled haploid plants by androgenesis of cucumber (*Cucumis sativus* L.). Plant Cell, Tissue and Organ Culture. 2007;**90**:245-254

[51] Baktemur G, Keles D, Kara E, Yildiz S, Taskin H. Effects of genotype and nutrient medium on obtaining haploid plants through ovary culture in cucumber. Molecular Biology Reports. 2022;**49**(6):5451-5458

[52] Nyirahabimana F, Solmaz I. Haploid induction through ovary culture in cucumber. In Vitro Cellular & Developmental Biology-Plant. 2024;**60**(1):122-130

[53] Domblides E, Shmykova N, Khimich G, Korotseva I, Kan L, Domblides A, et al. Production of doubled haploid plants of Cucurbitaceae family crops through unpollinated ovule culture in vitro. Acta Horticulture. 2020;**1294**:19-28. DOI: 10.17660/ActaHortic.2020.1294.4

[54] Sorntip A, Poolsawat O, Kativat C, Tantasawat PA. Gynogenesis and doubled haploid production from unpollinated ovary culture of cucumber (*Cucumis sativus* L.). Canadian Journal of Plant Science. 2017;**98**(2):353-361

[55] Parkash C, Bhatia R, Munshi AD, Rao M, Subhasmita S, Behera TK, et al. An efficient and easy-to-use protocol for induction of haploids in cucumber through parthenogenic embryo development. Botanical Studies. 2024;**65**(1):1-17

[56] Salehian H, Shahnazi S, Nazari M. Production of doubled haploid plants in cucumber (*Cucumis sativus* L.) via parthenogenesis. In Vitro Cellular & Developmental Biology-Plant. 2023;**59**(4):467-474

[57] Ebrahimzadeh H, Shariatpanahi ME, Ahmadi B, Soltanloo H, Lotfi M, Zarifi E.

Efficient parthenogenesis induction and in vitro haploid plant regeneration in cucumber (*Cucumis sativus* L.) using putrescine, spermidine, and cycocel. Journal of Plant Growth Regulation. 2018;**37**:1127-1134

[58] Li J, Si S, Cheng J, Li J, Liu J. Thidiazuron and silver nitrate enhanced gynogenesis of unfertilized ovule cultures of *Cucumis sativus*. Biologia Plantarum. 2013;**57**(1):164-168

[59] Khadija F, Fatima B, Usman M, Khan M. Floral bud size and culture conditions' effect on embryogenesis in anther-derived calli of cucumber. SABRAO Journal of Breeding and Genetics. 2024;**56**(5):2079-2089

[60] Amirian R, Hojati Z, Azadi P. Male flower induction significantly affects androgenesis in cucumber (*Cucumis sativus* L.). The Journal of Horticultural Science and Biotechnology. 2020;**95**(2):183-191

[61] Diao WP, Jia YY, Song H, Zhang XQ, Lou QF, Chen JF. Efficient embryo induction in cucumber ovary culture and homozygous identification of the regenetants using SSR markers. Scientia Horticulturae-Amsterdam. 2009;**119**(3):246-251

[62] Punja ZK, Tang FA, Sarmento GG. Isolation, culture and plantlet regeneration from cotyledon and mesophyll protoplasts of two pickling cucumber (*Cucumis sativus* L.) genotypes. Plant Cell Reports. 1990;**9**(2):61-64

[63] Cápal P, Ondřej V. Expression and epigenetic profile of protoplast cultures (*Cucumis sativus* L.). In Vitro Cellular & Developmental Biology - Plant. 2014;**50**(6):789-794

[64] Burza W, Malepszy S. In-vitro culture of Cucumis-Sativus L .18. Plants from protoplasts through direct somatic embryogenesis. Plant Cell, Tissue and Organ Culture. 1995;**41**(3):259-266

[65] Colijnhooymans CM, Bouwer R, Orczyk W, Dons JJM. Plant-regeneration from cucumber (*Cucumis sativus*) protoplasts. Plant Science. 1988;**57**(1):63-71

[66] Gajdová J, Navrátilová B, Smolná J, Lebeda A. Factors affecting protoplast isolation and cultivation of *Cucumis* spp. Journal of Applied Botany and Food Quality. 2012;**81**(1):1-6

[67] Jia SR, Fu YY, Lin Y. Embryogenesis and plant-regeneration from cotyledon protoplast culture of cucumber (*Cucumis sativus* L). Journal of Plant Physiology. 1986;**124**(5):393-398

[68] Petrivalsky M, Vanícková P, Ryzí M, Navrátilová B, Piterková J, Sedlárová M, et al. The effects of reactive nitrogen and oxygen species on the regeneration and growth of cucumber cells from isolated protoplasts. Plant Cell, Tissue and Organ Culture. 2012;**108**(2):237-249

[69] Moricova P, Ondrej V, Navratilova B, Luhova L. Changes of DNA methylation and hydroxymethylation in plant protoplast cultures. Acta Biochimica Polonica. 2013;**60**(1):33-36

[70] Fellner M, Lebeda A. Callus induction and protoplast isolation from tissues of *Cucumis sativus* L. and C-Melo L. seedlings. Biologia Plantarum. 1998;**41**(1):11-24

[71] Bairu MW, Aremu AO, Van Staden J. Somaclonal variation in plants: Causes and detection methods. Plant Growth Regulation. 2011;**63**(2):147-173

[72] Cassells AC, Curry RF. Oxidative stress and physiological, epigenetic and genetic variability in plant tissue culture:

Implications for micropropagators and genetic engineers. Plant Cell, Tissue and Organ Culture. 2001;**64**(2-3):145-157

[73] Park HS, Jeon JH, Cho W, Lee Y, Park JY, Kim J, et al. High-throughput discovery of plastid genes causing albino phenotypes in ornamental chimeric plants. Horticulture Research-England. 2023;**10**(1):uhac246

[74] Custers JBM, Zijlstra S, Jansen J. Somaclonal variation in cucumber (*Cucumis sativus* L) plants regenerated via embryogenesis. Acta Botanica Neerlandica. 1990;**39**(2):153-161

[75] Skarzynska A, Pawelkowicz M, Plader W. Genome-wide discovery of DNA variants in cucumber somaclonal lines. Gene. 2020;**736**:144412

[76] Plader W, Malepszy S, Burza W, Rusinowski Z. The relationship between the regeneration system and genetic variability in the cucumber (*Cucumis sativus* L.). Euphytica. 1998;**103**(1):9-15

[77] Ladyzynski M, Burza W, Malepszy S. Relationship between somaclonal variation and type of culture in cucumber. Euphytica. 2002;**125**(3):349-356

[78] Bartoszewski G, Havey MJ, Ziolkowska A, Dlugosz M, Malepszy S. The selection of mosaic (MSC) phenotype after passage of cucumber (*Cucumis sativus* L.) through cell culture: A method to obtain plant mitochondrial mutants. Journal of Applied Genetics. 2007;**48**(1):1-9

[79] Zhao ZY, Qi YG, Yang ZM, Cheng LY, Sharif R, Raza A, et al. Exploring the agrobacterium-mediated transformation with CRISPR/Cas9 in cucumber (*Cucumis sativus* L.). Molecular Biology Reports. 2022;**49**(12):11481-11490

[80] Xin TX, Tian HJ, Ma YL, Wang SH, Yang L, Li XT, et al. Targeted creation of new mutants with compact plant architecture using CRISPR/Cas9 genome editing by an optimized genetic transformation procedure in cucurbit plants. Horticulture Research-England. 2022;**9**:uhab086

[81] Nanasato Y, Konagaya K, Okuzaki A, Tsuda M, Tabei Y. Improvement of agrobacterium-mediated transformation of cucumber (*Cucumis sativus* L.) by combination of vacuum infiltration and co-cultivation on filter paper wicks. Plant Biotechnology Reports. 2013;**7**(3):267-276

[82] Nishibayashi S, Kaneko H, Hayakawa T. Transformation of cucumber (*Cucumis sativus* L) plants using agrobacterium tumefaciens and regeneration from hypocotyl explants. Plant Cell Reports. 1996;**15**(11):809-814

[83] Ni Z, Li J, Zhang S, Dong Y, Liu Q, Li M, et al. Construction and functional analysis of nattokinase-producing cucumber obtained by the CRISPR-Cas9 system. Acta Alimentaria Hung. 2023;**52**(1):121-131

[84] Tek MI, Calis O, Fidan H, Shah MD, Celik S, Wani SH. CRISPR/Cas9 based mlo-mediated resistance against Podosphaera xanthii in cucumber (*Cucumis sativus* L.). Frontiers in Plant Science. 2022;**13**:1081506

[85] Shnaider Y, Elad Y, Rav-David D, Pashkovsky E, Leibman D, Kravchik M, et al. Development of powdery mildew resistance in cucumber using CRISPR/Cas9-mediated mutagenesis of CsaMLO8. Phytopathology. 2023;**113**(5):786-790

[86] Fidan H, Calis O, Ari E, Atasayar A, Sarikaya P, Tek MI, et al. Knockout of

eIF4E using CRISPR/Cas9 for large-scale production of resistant cucumber cultivar against WMV, ZYMV, and PRSV. Frontiers in Plant Science. 2023;**14**:1143813

[87] Chandrasekaran J, Brumin M, Wolf D, Leibman D, Klap C, Pearlsman M, et al. Development of broad virus resistance in non-transgenic cucumber using CRISPR/Cas9 technology. Molecular Plant Pathology. 2016;**17**(7):1140-1153

[88] Zheng XY, Yang JJ, Lou TX, Zhang J, Yu WJ, Wen CL. Transcriptome profile analysis reveals that CsTCP14 induces susceptibility to foliage diseases in cucumber. International Journal of Molecular Sciences. 2019;**20**(10):2582

[89] Mitra D, Shnaider Y, Bar-Ziv A, Brotman Y, Perl-Treves R. First fruit inhibition and CRISPR-Cas9 inactivation of candidate genes to study the control of cucumber fruit set. Acta Horticulturae. 2020;**1294**:135-140

[90] Peng YQ, Chen LL, Zhu L, Cui LJ, Yang L, Wu HH, et al. Is a key gene for the CeO nanoparticle's improved cucumber salt tolerance: A validation from CRISPR-Cas9 lines. Environmental Science: Nano. 2022;**9**(12):4367-4381

[91] Dong SY, Li CX, Tian HJ, Wang WP, Yang XY, Beckles DM, et al. Natural variation in STAYGREEN contributes to low-temperature tolerance in cucumber. Journal of Integrative Plant Biology. 2023;**65**(12):2552-2568

[92] Nguyen DV, Hoang TTH, Le NT, Tran HT, Nguyen CX, Moon YH, et al. An efficient hairy root system for validation of plant transformation vector and CRISPR/Cas construct activities in cucumber (*Cucumis sativus* L.). Frontiers in Plant Science. 2022;**12**:770062

[93] Kiryushkin AS, Ilina EL, Kiikova TY, Pawlowski K, Demchenko KN. Do DEEPER ROOTING 1 homologs regulate the lateral root slope angle in cucumber (*Cucumis sativus*)? International Journal of Molecular Sciences. 2024;**25**(4):1975

Chapter 5

Representative Species of the Cucurbitaceae Family: Valuable Phytochemicals and Properties for Food and Medicine

Anna Busuioc, Bianca Furdui, Andreea-Veronica Botezatu and Rodica-Mihaela Dinică

Abstract

Recent research regarding various species from the Cucurbitaceae family, which are used in traditional medicine, has highlighted remarkable chemical composition and therapeutic properties with possible applications in modern alternative medicine and food field. The species mentioned in this chapter are *Cucumis metuliferus*, *Benincasa hispida*, and *Trichosantes cucumerina*, as well as a more studied and used species among them, *Momordica charantia*. This chapter correlates the phytochemical composition of these species (primary and secondary metabolites) with their biological potential, which confers therapeutic and nutritional properties. This chapter aims to highlight the importance of the four mentioned species belonging to the Cucurbitaceae family, along with their wide range of applications and also correlated with the different classes of phytochemical compounds.

Keywords: Cucurbitaceae family, phytochemical composition, therapeutic properties, food, representative species

1. Introduction

Since ancient times, people around the world have used plants as valuable sources of food and medicines [1]. In Europe, plant chemistry and the isolation of biologically active compounds developed rapidly after the influence of Paracelsus (sixteenth century), who claimed that "*All meadows and pastures, all mountains and hills are pharmacies*" [1], that is, nature is a supplier of natural products and medicines. After decades of research and evolution, scientific research is still exploring traditional medicine, and attempts are being made to elucidate the power and benefits that nature offers to integrate it into conventional medicine.

The world's biggest challenges, in recent years, have been the growth of the human population and the depletion of natural resources, which have led to hunger, malnutrition, and poor human health [2]. Other important aspects of today's life are the increase in the number of human diseases, the development of resistance of some diseases to classical treatments, as well as the toxic side effects of synthetic drugs [3, 4]. The Food and Agriculture Organization of the United Nations (FAO) estimated that approximately 9.1% of the global population suffered from chronic malnutrition in 2023. Most of the malnourished people live in low and sub-middle-income regions [5]. Africa has the highest prevalence of undernutrition, but Asia, being the most densely populated region of the world, has the highest percentage of undernourished people [6]. In the current conditions of climate change and the underlying impact on agriculture and horticulture, underutilized plants are valuable alternatives for cultivation and use for various purposes.

Nowadays, in agriculture, grains, vegetables, and fruits are produced by using traditional or genetically modified plants and modern agricultural practices, while attempts are made to adapt various plant species to new growing environments. However, there are many plant species still underutilized or underexploited, including plant species that thrive in extreme environmental conditions and threatened habitats, possessing genetic tolerance to survive in harsh conditions and that also have important nutritional, phytochemical, and therapeutic qualities [7]. Various research has made significant advancements in demonstrating that many compounds isolated from plants are biologically active and important in the prevention of chronic conditions such as cancer, diabetes, and coronary heart disease. Phytochemical compounds have demonstrated numerous bioactivities that include antioxidant, antitumoral, antimicrobial, antihypertensive, anti-inflammatory, and hepatoprotective properties [6, 8]. Vegetables from the Cucurbitaceae family are commonly known as cucumbers. The oldest species of this plant family, first described in 1924 by paleobotanist Frank Hall Knowlton and identified to date, is *Cucurbitaciphyllum lobatum* from the Paleocene epoch in Shirley Canal, Montana [9]. Cucumbers are believed to have their origin in India and some areas from Western Asia [10]. Some species used as natural remedies in traditional medicine grow in the foothills of the Himalayas, but they are not cultivated at large scale because of their bitter fruits, dormant seeds, or late maturity [9].

Other related species in the Cucurbitaceae family were cultivated in China and the Middle East. According to available data, originally from India, the cucumber was spread to Greece and Italy, where, through the Roman Empire, it was cultivated in other regions from Europe [8, 11]. Other records of cucumber cultivation were found in France in the ninth century and in Britain in the fourteenth century, while varieties more closely related to the pumpkin and cantaloupe were first cultivated in America, successfully cultivated later in Southern Mexico since 8000 BC, in the Illinois region around 5000 BC, and in Europe during the exploration of Columbus [12].

Cucurbitaceae is a diverse plant family currently cultivated worldwide in a variety of environmental conditions. Optimal cultivation conditions are tropical and subtropical, but it can be easily adapted to cultivation in other continents. Over 300 plant species are used by humans, but only 150 species are cultivated extensively, from which only 30 are crucial for global food production. The main producers of Cucurbitaceae are countries such as India, Turkey, China, and the United States [11]. From the multitude of known and used species, in this chapter, four species of therapeutic and nutritional interest have been selected: *Momordica charantia*, *Benincasa hispida*, *Trichosantes cucumerina*, and *Cucumis metuliferus*.

The Cucurbitaceae family has an important economic role because the vegetables and fruits of this family are used in various food, cosmetic or pharmaceutical products, or for other purposes (**Figure 1**). Edible plants, derived from the Cucurbitaceae

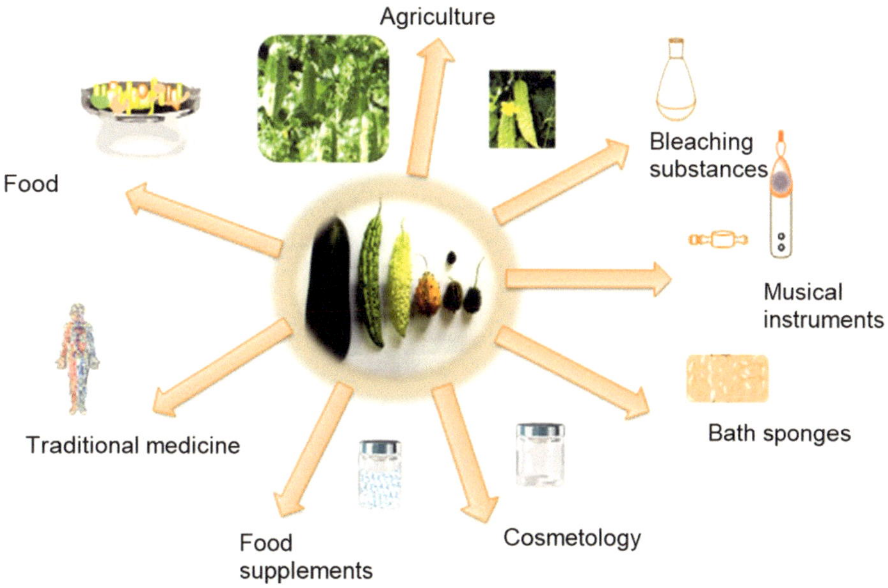

Figure 1.
Various uses of plant species from Cucurbitaceae family.

family, belong to a group that includes five genera: *Cucurbita*, *Cucumis*, *Citrullus*, *Legenaria*, and *Sechium* [13]. This family with characteristic seeds, peel deposits, a specific chemical composition, and diverse applications, can be easily identified throughout the plant kingdom [14, 15].

2. Important classes of organic compounds identified in the studied plants

Plants can be compared to "factories" that, under the action of solar energy, consume CO_2 and water and then produce oxygen, energy, and basic compounds, such as proteins, carbohydrates, vitamins, and lipids, which are necessary for the survival of living organisms [16]. Plant metabolism consists of numerous chemical reactions catalyzed by enzymes, which take place in various plant cell structures and the production of a variety of chemical molecules from the category of primary metabolites [17]. Primary metabolites play a major role in maintaining the viability of the plant (proteins, sugars, vitamins, and lipids). In addition, plants also synthesize a series of other compounds in the chemical processes of secondary metabolism such as terpenes, phenolic, nitrogen (alkaloids, cyanogenic glycosides, and nonprotein amino acids), and sulfur compounds [18]. Secondary metabolites are biosynthesized by plants and specific to each species, often manifesting functions characteristic of some plant organs or tissues; can be identified only in a certain period of growth and development within a species; or can be activated only under certain stress factors such as the attack of microorganisms, insects or the lack of nutrients [19]. Secondary metabolites are considered compounds of interest for various reasons, such as their structural diversity and their potential as drug candidates [18].

In general, secondary metabolites are produced by biosynthetic modifications of primary metabolites, including methylation, glycosylation, and hydroxylation. Plants contain highly specific secondary metabolites, sometimes toxic at high concentration, and represent a valuable library of bioactive compounds with broad activity on human cells, bacteria, fungi, and parasites [20].

2.1 Primary metabolites

The main primary metabolites, namely, saccharides, lipids, proteins, and vitamins, and their distribution in plant species from the Cucurbitaceae family are briefly characterized. Primary metabolites fulfill essential metabolic roles by participating in plant growth, nutrition, and reproduction.

2.1.1 Saccharides

Saccharides, also known as carbohydrates or sugars, are natural compounds synthesized by plants in the process of photosynthesis [15, 21]. It is known that saccharides fulfill various functions in plants, among them being energy sources for vital processes and important precursors in the biosynthesis of other chemical molecules [22]. This class of compounds acts as nutrients, plays an important role as signaling molecules, and regulates vital functions, that is, growth processes, metabolism, and stress tolerance response [23]. Low molecular mass saccharides, such as sucrose, trehalose, or raffinose, are involved in plant immunity as signaling molecules, which has led to the concept of "sweet immunity" [24]. These compounds are able to activate defenses and resistance-enhancing responses to pathogens [25]. Polysaccharides are widespread in plants and perform various functions such as structural (cellulose), nutritional reservoir (starch), or specific roles such as anticoagulant (heparin) [26].

In the fruits of *Momordica charantia*, various mono- and disaccharides such as arabinose, galactose, glucose, mannose, ribose, xylose, and rhamnose have been identified [27]. Studies carried out on the species of *M. charantia* from Nigeria have shown that, depending on the part of the analyzed plant, the carbohydrate content differs. Thus, similar values were found in leaves, 32.34 ± 0.24%, and fruits, 34.31 ± 0.30%, close also to those obtained from another study of 39 ± 3% [28], while the seeds were twice as poor in sugars with 9.18 ± 0.86% [29]. Analyses carried out on the seeds of a local variety of *M. charantia* collected from an agricultural research institute in Pakistan presented a higher concentration of soluble carbohydrates, 28.43 ± 3.11%, compared to previous studies [30]. In fresh *M. charantia* fruits from the local market in Penang, Malaysia, a percentage of 45.4 ± 0.9% maltose equivalents was quantified, while in freeze-dried fruits the following saccharides and their derivatives were identified: glucose (0.383 ± 0.004%), galactose (0.306 ± 0.002%), mannose (0.010 ± 0.001%), galacturonic acid (0.145 ± 0.003%), rhamnose (0.021 ± 0.000%), arabinose (0.086 ± 0.000%), xylose (0.049 ± 0.000%), as well as very small concentrations of ribose and fructose [31].

Fresh fruits of *Cucumis metuliferus* have been found to have a low carbohydrate content of 5.55 g/100 g fresh fruit [32], and in dried fruits, 7.56 g/100 g fruit [33] or in another study, 62.25 + 1.95 mg/kg [34]. Studies on aqueous extracts from the peel of fruits grown in China have shown the predominant presence of galacturonic acid, 45.93%, with higher amounts of rhamnose (13.50%), arabinose (10.37%), glucose (5.54%), and smaller amounts of fructose (2.51%), xylose (4.24%), and mannose (0.17%) [35].

Although carbohydrates play an important role in a wide range of biological processes, carbohydrate-derived drugs cover only a limited area of therapeutic actions [36]. Many pathophysiological important carbohydrate-protein interactions have not yet been exploited due to their high polarity. Saccharides cannot passively cross the enterocyte layer of the small intestine (a prerequisite for oral availability), and if systemically available by parenteral administration, these compounds undergo a process of rapid renal excretion. In addition to a lack of affinity, saccharides present low tissue permeability, short plasma half-life, and poor stability [36].

Medicines with compounds derived from saccharides in their structure, currently approved for human use, include benefits such as the treatment of diabetes mellitus due by the inhibition of glycosidase that prevent the digestion of carbohydrates (voglibose, miglitol and acarbose), the prevention of influenza virus infections (zanamivir and oseltamivir), anticoagulant properties by binding to antithrombin III for the treatment of sulfated glycosaminoglycan thrombosis (fondaparinux, daltepa-rin, ardeparin, nardoparin and enoxaparin), treatment of Gaucher disease (miglu-stat), epilepsy (topiramate), and osteoarthritis (sodium hyaluronate) (**Figure 2**) [36–38]. As examples of saccharide-derived drugs emerge, the strategies developed for their design will pave the way to harness the potential of this class of compounds in various therapies [39]. In addition to their therapeutic potential, sugars are present in most food products.

2.1.2 Lipids

A class of compounds that include glycerides, phospholipids, waxes, or fatty acids, lipids play an important role in living organisms, being an essential com-ponent of the homeostatic function of the human body. Lipids contribute to some vital processes, having main biological functions in plants and human health [40].

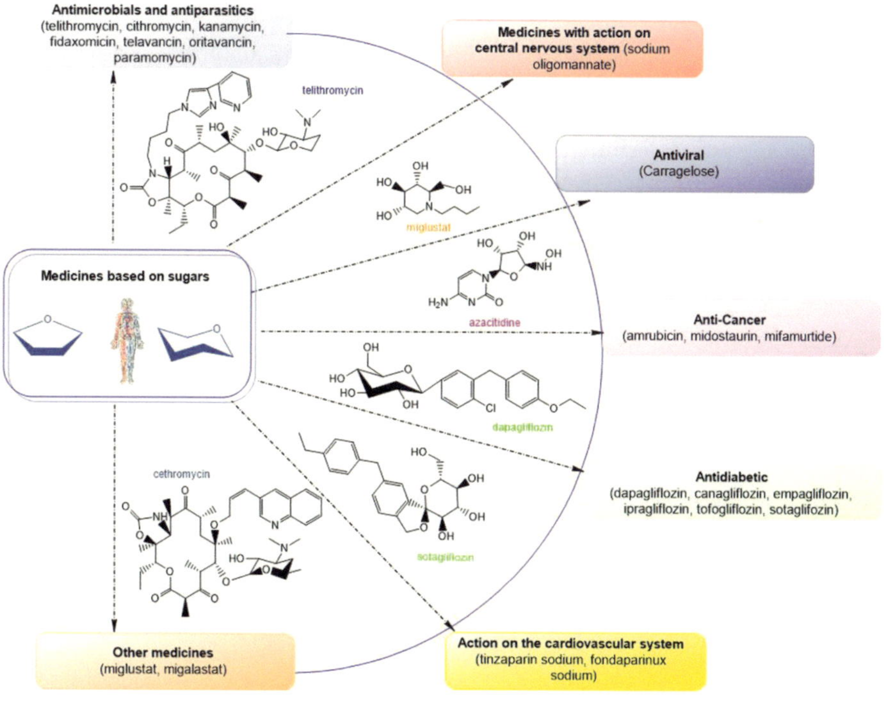

Figure 2.
Various applications of carbohydrates from plants.

Plants contain various types of lipids, including fatty acids, phospholipids, glycolipids, sterol lipids, sphingolipids, and waxes [41]. Lipids are usually extracted with organic solvents, which allow the removal of many hydrophilic components such as carbohydrates and proteins.

In higher plants, lipids are distributed in various parts, such as extrachloroplasts, where phosphatidylcholine and phosphatidylethanolamine are in higher concentrations, while thylakoids from chloroplasts usually contain three glycosylglycerides (monogalactosyl, digalactosyl, and sulfoquinovosyl diacylglycerols) and phosphatidylglycerol. The acyl-lipid composition of higher plant tissues depends on their metabolism, the plant species, and cultivation environment [42].

The seeds of buffalo gourd (*Cucurbita foetidissima*), from the Cucurbitaceae family, that grows in the semiarid regions of the USA, Mexico, Lebanon, and India, contain a high amount of fats (32–39%) and also proteins [43]. Pumpkin seeds (*Cucurbita pepo*) are rich in linoleic acid, as well as in palmitic, stearic, oleic acid, and a wide range of other fatty acids in moderate concentrations [44]. Their chemical composition has attracted attention due to its beneficial properties in the early treatment of prostate cancer. Microwave-assisted and supercritical carbon dioxide extraction revealed the triacylglycerols, sterols, and tocopherols due to which the seeds have this beneficial effect [45, 46].

Benincasa hispida seeds presented different content of saturated and unsaturated fatty acids such as myristic acid, palmitoleic acid, oleic acid, linoleic acid, stearic acid, α-linolenic acid, palmitic acid, and others [47]. In one of the more detailed studies in which a supercritical fluid extraction method was optimized, the total content of saturated fatty acids (myristic acid, palmitic acid, and stearic acid) was 15.97%, compared to palmitic acid, which was identified in a higher concentration of 9.83 ± 0.33%. Unsaturated fatty acids were determined in the highest concentration in percentage of 83.53% of the total fatty acids, these being represented by palmitoleic acid, oleic acid, linoleic acid, and linolenic acid, especially α-linolenic acid, an omega-3 type acid [48]. In addition to fatty acids, sterols such as β-sitosterol and stigmast-5-ene-3-p-ol have also been identified [47]. In the edible part of the winter cucumber, a fat content of 0.02–0.20/100 g was determined [49], a content of 0.60 ± 0.10/100 g in the pulp, and 1.25 ± 0.06/100 g [50] in the peel.

The highest content of lipids is mostly found in the seeds of plants. Thus, in bitter cucumber, about 64.3% are polyunsaturated fatty acids and 20.1% are monounsaturated fatty acids of the total fatty acids. The oil of these seeds is unique due to its high content of conjugated 9-cis, 11-trans, 13-trans linolenic acid [51]. Another study demonstrated high concentrations of stearic acid (62.31% of total fatty acids), oleic acid (12.53%), and linoleic acid (10.40%), as well as in smaller concentrations arachidic acid (5.3%), palmitic acid (4.72%), capric acid (0.53%), and lauric acid (0.45%) [49].

In the case of kiwano fruits, the seeds showed the highest lipid concentration, 15.4 g/100 g, followed by the fruit peel, 8.89 g/100 g, and a small content in the fresh pulp [52]. Another study showed a higher value for lipid and protein in seeds (23.8 and 23.2 g/100 g, respectively) [53]. The seeds contain fatty acids in different proportions, such as linoleic acid (56.21%) and oleic acid (18.6%), followed by palmitic acid (16.2%) and stearic acid (7.65%) [54, 55].

It has been demonstrated that the chemical composition of fresh and dried fruits differs. Thus, fresh *Cucumis metuliferus* fruits presented a low lipid content of 0.7 g/100 g [8] and the dried ones of 1.26 g/100 g fruit [56]. In another study, a lipid content of 7.8 ± 0.76 mg/kg was determined for dried fruits [57]. The scientific literature illustrates the fact that lipids are also found in the leaves of some cucumber species, with 2.6 ± 0.30 mg/kg, these values being higher than those found in the fresh fruit [31, 58].

The fruits of *Trichosanthes cucumerina* presented a small percentage of lipids, 0.47 ± 0.03% [59]. Instead, a special composition was identified in seeds of this plant, its oil containing medium concentrations of octanoic acid, dodecanoic acid, tetradecanoic acid, n-hexadecanoic acid, oleic acid, and octadecanoic acid [14]. In another study, a different composition was found rich in lauric acid (10.28 ± 0.01%), myristic acid (6.54 ± 0.01%), palmitic acid (11.87 ± 0.01%), stearic acid (15.54 ± 0.01%), oleic acid (26.51 ± 0.01%), linoleic acid (0.75 ± 0.01%), and palmitoleic acid (0.45 ± 0.01%) [60].

Global awareness of the concept of studying a group of molecules with lipid structures with therapeutic and prophylactic applications in promoting health and treating various ailments, supported by progressive research efforts to identify the properties and potential applications of such bioactive food constituents and together with consumer demand, has+ led to the development of a new food category commonly referred to as "functional foods" or "nutraceuticals" [42, 61]. Among the multitude of promising functional lipids that are known to date, it should be noted that the scope of functional lipids involves both bioactive lipids of natural and synthetic origin [62]. Functional lipid supplementation in the human diet can act as either an alternative therapy or an adjunctive therapy to conventional medicinal practices. Their importance increases in the current context, because nowadays, conventional therapies have many side effects and the diseases are more and more difficult to treat [42, 63]. Species in the Cucurbitaceae family have a varied lipid content, which can be useful in preventing or combating certain conditions.

2.1.3 Proteins

Proteins are another class of primary metabolites that are indispensable for living organisms. Proteins are polypeptides with a molecular weight greater than 10,000, which are made up of four basic elements, namely, carbon 50–55%, oxygen 20–23%, nitrogen 12–19%, and hydrogen 6–7%, as well as other elements in smaller amounts such as phosphorus, sulfur, copper, zinc, and others [14]. The main sources of proteins are from animals, but they are also found in smaller amounts in plants, being considered incomplete or low-quality proteins because their structures lack some of the essential amino acids. Proteins from oilseeds and legumes are generally soluble in aqueous media, and those from cereals are insoluble; therefore, it is necessary to use suitable extraction solvents. The separation of proteins is a difficult process due to their sensitivity, which takes into account temperature, degree of ionization, as well as extreme pH values, which can cause protein denaturation [64].

The protein content for the fruit of *Cucumis metuliferus* was found to be between 1.78 and 1.80 g/100 g, for the peel 2.95 g/100 g, and for the seeds, 2.63 g/100 g [56, 65, 66]. A study carried out on plant materials from Nigeria, Africa, presented a protein content in fruits of 9.04 ± 0.13 mg/kg, and of 9.77 ± 0.30 mg/kg in leaves [31]. Fresh fruits have a high concentration of proteins, of about 4 g/100 g [67]. The most recent studies highlight this class of compounds in kiwano fruits, varying between 6.22% and 6.29% [68].

The preliminary phytochemical screening of the extract from the dried seeds of *Benincasa hispida* highlights, in addition to the presence of carbohydrates, phenolic compounds, amino acids, flavonoids and sterols, the presence of proteins [56], such as trigonelline and osmorin, known for their antibacterial properties [69]. The highest protein content was found in the seeds of winter cucumber, 0.58 ± 0.02 g/100 g, followed by the edible part of the plant, 0.49 ± 0.02 g/100 g, and finally the peel of the fruit, 0.37 ± 0.03 g/100 g [50]. Studies in the literature have shown that *Trichosanthes cucumerina* seeds have a rich protein content of 26.2–26.6 g/100 g [70]. *Momordica charantia* fruits showed a protein content of 0.71 ± 0.13% [14], including polypeptide-P, lectin, and MAP30 with anti-HIV properties [64, 71]. Proteins have also been isolated from the seeds of *M. charantia* [72–74]. In addition, the crude protein content of the fruit was found to be 11.4–20.9 g/kg, which is higher than that of ripe or common cucumbers [75]. Although it has many culinary uses, especially in South, Southeast, and East Asia, and is also cultivated for ornamental purposes, *M. charantia* is widespread in traditional medicine [76]. Fruits, flowers, and young shoots are also used as flavoring agents in various Asian dishes. The young shoots and leaves of *M. charantia*, as well as leaf and fruit extracts, are used in the preparation of teas and dietary supplements [77]. The protein content and implicitly amino acid composition of plants vary considerably by family, genus, species, and cultivar. The role of some proteins in plants includes storage functions for lectins, enzymes, and enzyme inhibitors, and structural functions for ribosomal, chromosomal, and membrane proteins [78].

Proteins are considered valuable molecules due to their applications in the fields of food and medicine. Recently, therapeutic proteins have emerged as a promising option in treating various diseases, such as diabetes, cancer, cardiovascular diseases, and others. Finally, proteins can be used in combination with small molecular weight drugs to provide synergistic benefits in modern medicine [79].

2.1.4 Vitamins

Vitamins are considered primary metabolites of plants, being able to be further used in the synthesis of secondary metabolites [14, 80]. In general, vitamin biosynthetic pathways have been elucidated in various research [81, 82]. Vitamins, whether water-soluble or fat-soluble, play an important role in supporting the health of the human body by preventing and treating various conditions such as bone homeostasis, cancer, or other diseases of the immune, digestive, skin, and cardiovascular systems [14, 81, 83–85].

Vitamins are compounds sensitive to different factors such as heat, light, oxygen, and the pH of the environment in which they are stored. There are 13 compounds recognized so far as vitamins that have been detected in plants, which are important for living organisms. Thus, fruits and vegetables, as well as various cereals and nuts, can provide in the human diet approximately 92% ascorbic acid, 69% folic acid, 64% thiamine, 49% vitamin A, 49% niacin, 46% vitamin B6, and 33% riboflavin [81]. The concentration of vitamins differs considerably in the leaves, fruits, stems, and roots of plants. It has been observed that certain parts of plant tissues located toward the outside usually contain higher concentrations of vitamins, especially ascorbic acid. Cucumbers, for example, contain a higher concentration of ascorbic acid, riboflavin, niacin, pantothenic acid, and tocopherol in the skin than in the inner part of the fruit [84].

Momordica charantia fruits are rich in various vitamins of major importance, especially vitamin C [86, 87], vitamin A, vitamin B complex [75], and vitamin E [88]. Vitamin A, vitamin E, vitamin C, vitamin B12, and folic acid have been identified in small amounts in the dried leaves of this plant, and very small concentrations of vitamin B3, vitamin B6, vitamin D, and vitamin K [89]. The main vitamin identified from the pulp of kiwano fruits was vitamin C in variable concentration of 15.5–27.5 g/100 g dry matter or, according to other studies, 5.3 mg/g dry matter [50], followed by vitamin E (13.5–8.3 g/100 g dry matter), folic acid (0.003 mg/g dry matter), vitamin A (0.007 mg/g dry matter), riboflavin (0.015 mg/g dry matter), thiamine (0.025 mg/g dry matter), niacin (0.56 mg/g dry matter), and pantothenic acid (0.18 mg/g dry matter) [90]. Vitamin C was found in large amounts of 625.12 ± 11.17 mg/100 g [50] and vitamin K in a proportion of 1.69 mg/g dry matter from the peel of *Cucumis metuliferus* [84]. This content is higher than the recommended daily intake, which is from 70 to 140 µg for an adult [87]. Vitamins B1 were quantified in the peel of *C. metuliferus* at a concentration of 1.69 mg/g dry matter, along with vitamin B9 (1.93 mg/g dry matter) and vitamin B2 (1.74 mg/g dry matter), vitamins D (2.28 mg/g dry matter), and vitamin E (2.92 mg/g dry matter), vitamins that were not identified in the pulp of the same species. Fat-soluble vitamins were found in the seeds, such as vitamins D (1.86 mg/g dry matter), vitamin E (2.05 mg/g dry matter), vitamin K (2.58 mg/g dry matter), and vitamin A (2.22 mg/g dry matter). Among the water-soluble vitamins, the complex of B vitamins was identified (thiamine 2.20 mg/g dry matter, folic acid 2.20 mg/g dry matter, and riboflavin 1.72 mg/g dry matter), these being present in higher concentrations than vitamin C (1.72 mg/g dry matter), vitamin that was not identified in seeds [50].

Vitamin C (1.35–0.68 mg/g dry matter), thiamin (0.02–0.04 mg/g dry matter), riboflavin (0.02–0.31 mg/g dry matter), niacin (0.20–0.46 mg/g dry matter), and vitamin E were identified in *Benincasa hispida* [14, 69]. A comparative study of fruits from different countries demonstrated the presence of vitamins in different proportions; thus, the highest concentrations of vitamin C (68 mg/100 g dry matter), thiamine (0.02 mg/100 g dry matter), riboflavin (0.31 mg/100 g dry matter), and niacin (0.2 mg/100 g dry matter) were quantified in plants from Malaysia, while those from Australia had a different composition of 27 mg/100 g dry matter of vitamin C, 0.4 mg/100 g dry matter of niacin, 0.05 mg/100 g dry matter of riboflavin, and 0.02 mg/100 g dry matter of thiamine. Another study on *Benincasa hispida* from China presented the poorest vitamin content of 1.35 mg vitamin C/100 g plant, 0.02 mg riboflavin/100 g, 0.46 mg niacin/100 g, and a lack of thiamine [91].

A comparative study highlighted the fact that chemical composition is influenced by the harvesting period of plants. Thus, it was demonstrated that the highest concentration of vitamin C in the fruits of *T. cucumerina* is found in the late harvest period (0.94 g/100 g), being higher than in the crops at the beginning of the season (0.252 g/100 g) [11, 92]. In another study in which the edible part of the fruit was analyzed, 30.5% vitamin C, 11.3% vitamin B6, 3% pantothenic acid, and less than 1.1% of vitamin E were present, while vitamins B12 and D were not identified [93].

Finally, vitamins are indispensable in the functioning processes of the human body, and the lack of vitamins leads to the impairment of metabolism. Some vitamins are very unstable and can be easily lost, especially in food processing and storage, so processing methods and storage conditions are very important to maintain vitamin content. In addition, the content and variety of vitamins in food are different, and sometimes vitamin supplements are recommended to meet the needs of the human body [73].

2.2 Secondary metabolites

Secondary metabolites are important compounds in plant growth and development, with numerous applications in various industrial, food, and therapeutic fields, and have been increasingly researched in recent years to discover potential therapeutic targets [16]. There are various classifications of these secondary metabolites. Some of these classify secondary metabolites in pigments, glycosides, lignins, tannins, resins, balms, alkaloids, antibiotics, phytoncides, vegetable insecticides, and others [24]. The presence of secondary metabolites in plants contributes to their pollination, which is carried out with the help of insects and is due to different classes of compounds that influence the color, fragrance, or nectar of these plants [86, 94].

In general, secondary metabolites are generated by biosynthetic modifications of primary metabolites. For economic reasons, the study of secondary metabolites has been the subject of research in recent years, contributing to the development of several fields of phytochemistry. Studies have identified over a 1000 new chemical compounds from different plant species annually. According to literature data, approximately 25% of drugs used to prevent and treat various ailments contain compounds of plant origin [18, 20, 95].

Secondary metabolites have important roles in plants as biologically, physiologically, and ecologically active compounds, but they are also important in various domains such as food, pharmaceutical, technical, industrial, or agricultural fields.

2.2.1 Polyphenols

Phenolic compounds are secondary metabolites of plants, characterized by the presence of an aromatic ring and several hydroxyl groups. Probably the largest group of secondary metabolites, phenolic compounds range from simple (with a single aromatic ring) to complex structures. Their primary role in plants is to protect the plant from being attacked by pathogens, free radicals, or UV rays [96].

According to specialized studies, *Benincasa hispida* has an increased content of flavonoids in the peel, 1.13 ± 0.15 g/100 g, while in the seeds, they were not identified [44]. The contents of phenolic acids and flavonoids in the dried fruit were 81.3 ± 1.4 mg GAE/g and 486.8 ± 4.0 mg CE/g dry product, respectively [97]. By another method of isolation and investigation, values of 193.77 ± 2.86 µg in 10 mg of aqueous extract were determined [98].

In *Momordica charantia*, flavonoid compounds were found in two forms, namely, as free aglycones or glycosides. Depending on the type of extract, the concentration of polyphenols and flavonoids was found to vary, with the ethanolic and aqueous extracts showing a content of 44.0 mg and 62.0 mg EqQ/g extract, respectively, while the concentration of polyphenols was 68.8 and 51.6 mg EqAG/g extract [99]. The main phenolic compounds isolated from bitter cucumber were gallic acid, chlorogenic acid, caffeic acid, p-coumaric acid, syringic acid, vanillic acid, p-methoxybenzoic acid, protocatechuic acid, catechin, rutin, and myricetin. Glycosides such as apigenin-7-O-glycoside (1955.55 ng/mg) and luteolin-7-O-glycoside (725.50 ng/mg) were also identified [100], and other hydroxy acids such as ferulic, malic, malonic, salicylic, tartaric, and cinnamic acids. Phenolic acids such as caffeic, *o*-, *m*-, and *p*-coumaric acids were isolated from the roots of the plant, too [101, 102]. Other studies revealed chlorogenic, ferulic, p-hydroxybenzoic, gallic, β-resorcylic, vanillic, syringic, salicylic, cinnamic, veratric, homogentisic acids, and vanillin, as well as flavonoids such as myricetin, kaempferol, quercetin, catechin, rutin, hesperidin, naringenin, biocyanin A, and naringin, in *Momordica charantia* [103]. The leaves have a different phytochemical profile, which contain polyphenols such as 4-hydroxybenzoic acid, derivatives of 4-O-caffeoylquinic acid, 4-O-feruloylquinic acid, 5-O-feroylquinic acid, sinapinic acid, 2,4-bis(2-phenylpropan-2-yl)phenol and the acids found in other parts of the plant (caffeic acid, cinnamic acid, ferulic acid, and p-coumaric acid) [104], and flavonoids (isorhamnetin-3-O-glucoside, isorhamnetin-O-acetylhexoside, kaempferol-3-O-glucoside, kaempferol-3-O-rutinoside, kaempferol-O-acetylhexoside, kaempferol-O-pentosylhexoside, quercetin-3-O-glucoside, quercetin-3-O-rutinoside, quercetin-O-acetylhexoside, quercetin-O-dihexoside, quercetin-O-pentosylhexoside, and rutin) [105, 106]. The flowers, the less studied part of these plants, contain flavonoids that are represented by kaempferol and rutin and polyphenols such as 4-hydroxybenzoic acid, caffeic acid, epicatechin, catechin, chlorogenic acid, ferulic acid, gallic acid, and p-coumaric acid [107]. Studies in the literature showed that the seeds contain some phenolic compounds known and isolated in plants from other geographical areas, such as gallic acid, caffeic acid, epicatechin, catechin, chlorogenic acid, gentisic acid, o-coumaric acid, p-coumaric acid, sinapic acid, protocatechuic acid, syringic acid, cinnamic acid, ferulic acid, and vanillic acid [107, 108].

The presence of polyphenols and flavonoids in *T. cucumerina* fruit was initially highlighted by specific identification reactions. Subsequently, secondary metabolite contents were determined by spectrophotometric and chromatographic methods. Thus, from the most recent studies, the total polyphenol content was between 3.1 and 5.8 EqAG/g dry matter [90]. The peels of the fruit are rich in both polyphenols, 500 mg EqAG/100 g dry matter, and flavonoids, 171 mg EqQ/100 g [109]. The compounds isolated from the pulp were rutin, quercetin, quercetin-3-d-galactoside, kaempferol-3-glucoside, kaempferol [109], gallic acid, catechin, epicatechin, neo-chlorogenic acid, caffeic acid, and p-coumaric acid [90]. The studies that were carried out on the aerial part of *T. cucumerina* plants showed that they have a total content of polyphenols and flavonoids of 38.82 ± 0.45 mg EqAG/g extract and 26.30 ± 0.45 mg EqQ/g extract, respectively.

Phytochemical screening of the studied plant materials from the Western Province of Sri Lanka highlighted the presence of polyphenols, flavonoids, steroids, and tannins [110]. A total content of 4.64 ± 0.3 mg EqAG/g of polyphenols and 0.77 ± 0.1 mg Eq. Q/g of flavonoids was determined for the fruits of the snake cucumber. A higher content was determined in the plant's flowers (19.97 ± 1.6 mg EqAG/g and 4.46 ± 0.1 mg Eq. Q/g), while the richest content was in the leaves, which presented 27.39 ± 1.2 mg EqAG/g polyphenols and 6.05 ± 0.1 mg Eq. Q/g flavonoids [111]. For the aqueous extracts of the ripe fruits, 65.8 ± 0.50 mg EqAG/g total polyphenols and 45.4 ± 0.76 mg Eq. Q/g total flavonoids were quantified [112].

Polyphenols detected in species of the Cucurbitaceae family play a crucial role due to their potent antioxidant, antimicrobial, and anti-inflammatory activities, which contribute to reducing oxidative stress, modulating cell signaling pathways and preventing the onset and progression of various chronic diseases, including cardiovascular diseases, neurodegenerative diseases, and certain types of cancer.

2.2.2 Alkaloids

Alkaloids represent a large group of nitrogen compounds with a basic character, which are found in plants as secondary metabolites, with more than 40,000 such compounds identified (**Figure 3**) [113].

Alkaloids play important roles in plant metabolism, catabolism, or plant physiology. They can act as nitrogen storage reservoirs, are protective agents against predator attack, growth regulators, or substitutes for plant minerals such as potassium and calcium [114].

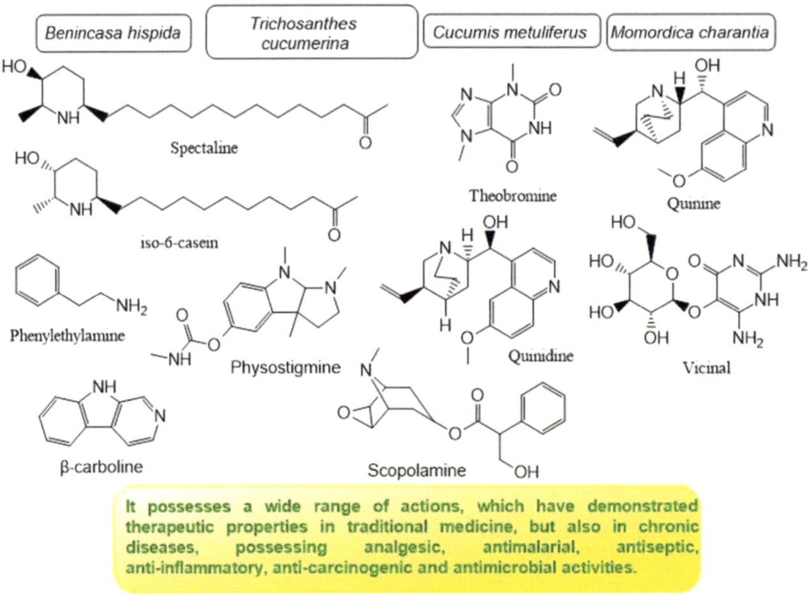

Figure 3.
Alkaloids with biological properties identified in Cucurbitaceae family.

Cucumis metuliferus has been found to be a rich source of alkaloids, at a concentration of 106 mg/g in the fruit peel, 1.06 mg/g in the bark, and up to 2.54 mg/g in the seeds [90].

In the *Benincasa hispida* species, alkaloids such as 5-methylcytosine and citrulline were identified, with a total content in the peel of 4.20 ± 0.10 mg/100 g, in the pulp 4.35 ± 0.05 mg/100 g, and no compounds of this class were identified in the seeds [44].

High concentrations of vicin were quantified in *Momordica charantia* fruits (0.210 ± 0.010 pg/100 pg), in fruit pulp (0.131 ± 0.005 pg/100 pg), and in peel (0.114 ± 0.006 pg/100 pg) [115, 116].

Some alkaloids have antibacterial, antifungal and antiviral, and anticancer properties. They also help reduce blood pressure, relieve pain and spasm, and stimulate circulation and breathing. These properties may extend to toxicity to animals, microorganisms, and viruses, and administered in an appropriate dose, some alkaloids are pharmacologically active [117].

Some alkaloids, such as spectaline, theobromine, quinine, quinidine, scopolamine, and physostigmine, with biological properties identified in the Cucurbitaceae family [70, 118–120] are presented in **Figure 3**.

The therapeutic properties of alkaloids are important in the current context, in which diseases are increasingly numerous and resistant to classical treatments. Therefore, additional research will be generated to discover new drugs based on natural compounds with reduced side effects compared to classical drugs.

2.2.3 Terpenes

Terpenes, as well as their oxygenated derivatives (terpenoids), are important plant metabolites. More than 40,000 terpenoids that have been identified share a common structural feature of being obtained by the repetitive fusion of branched five-carbon units based on the isoprene backbone. These monomers are generally referred to as isoprene units because the thermal decomposition of many terpenoid substances produces isoprene as a product [121].

Several compounds specific to this species have been isolated from the fruits and leaves of *M. charantia*, many of them being identified only in this plant species. Thus, until now, almost 204 triterpenoids have been isolated and identified from different parts of the *M. charantia* plant, most of which belong to the cucurbitan and oleanane-type terpenes [122]. Some of the most representative compounds from the terpenoid class isolated from the roots, flowers, and leaves of the *Momordica charantia* are α-carotene, lutein, β-carotene, β-ryptoxanthin, zeaxanthin, antheraxanthin, violaxanthin, squalene, and amyrin [123, 124]. Kuguacina P, a compound that has so far only been identified and isolated in bitter cucumber, was isolated from leaves in a proportion of 66.7%, in fruits 25.0%, and a smaller percentage in roots 8.3%. It was obtained by extraction using two basic solvents, ethanol with a yield of 58.3% and methanol with a yield of 25.0% [125].

A valuable compound from the terpenoid class that is more widespread in Cucurbitaceae species is carotene. Thus, the presence of β-carotene was identified in various parts of the species *Cucumis metuliferus* in the seeds in a proportion of 0.13 ± 0.06 mg/100 g seeds [54]. According to other studies, fruits contain between 1.5 and 88 μg β-carotene/100 g of fruit [90]. Other isolated compounds from *Cucumis metuliferus* plants were isovitexin, cucurbitacin C, lutein [126], cucurbitacin B [127], oleanolic acid, and ursolic acid [128].

T. cucumerina contains, in addition to β-carotene, lutein, and cucurbitacins, other compounds such as 23, 24-dihydrocucurbitacin D, 3-β-hydroxyoleic acid identified in the fruit, isocucurbitacin B, 23,24-dihydroisocucurbitacin B, 23,24-dihydrocucurbitacin E, and lycopene [129]. A comparative study on fresh and dry samples of plant material highlighted the presence of some compounds in different proportions, such as lycopene in the fresh pulp (18 mg/100 g), and in the dry pulp (165.3 mg/100 g), lutein (18.4 mg/100 g dry plant), α-carotene (10.7 mg/100 g dry plant), and β-carotene (2.8 mg/100 g dry plant) [130]. The dry plant material contains various concentrations of the same compounds (191.7 mg lutein/100 g, 101.5 mg α-carotene/100 g, and 29.2 mg β-carotene/100 g, respectively) [130]. Cucurbitacin B [131] and cucurbitacin E have been identified in the seeds of these species, too. Terpenoids specific to the Cucurbitaceae family, such as isornultiflorenol and cucurbitacin B, have been identified in the seeds of *B. hispida* [48].

Terpenes, a widespread class of natural plant compounds found abundantly in species of the Cucurbitaceae family, are of major importance for human health and well-being due to their diverse properties, including anti-inflammatory, antimicrobial, antioxidant, and antitumor activities. In addition to their therapeutic role, terpenes are the compounds that contribute to the aroma, taste, and smell of plants, having valuable applications in the food, pharmaceutical, cosmetic, and aromatherapy industries, thus strengthening the essential link between plants and the quality of human life.

2.2.4 Saponins

Saponins are a heterogeneous group of glycosides that contain at least one glycosidic bond and are characterized by their structure containing a triterpene or steroid aglycone and one or more monosaccharide units. They are widespread in the plant kingdom, especially in legumes [132]. These compounds are secondary plant metabolites that are stored in plant cells as inactive precursors but can be easily transformed into biologically active substances by plant enzymes in response to pathogen attack.

In several studies, the presence of saponins in *C. metuliferus* fruits is demonstrated, but their chemical composition has not been studied in detail [133].

Phytochemical screening on *T. cucumerina* species revealed the presence of various classes of compounds, including saponins [90].

Extracts from *M. charantia* seeds presented saponins such as goiaglycoside A, goiaglycoside B, goiaglycoside C, goiaglycoside D, momordicoside F2, momordicoside I, and momordicoside K [76].

Consumer demand for natural products, together with their physicochemical (surfactant) properties and increasing evidence of their biological activity, has led to the study of saponins as commercially significant compounds with expanding applications in the food, cosmetic, and pharmaceutical fields. Increasing their commercial potential requires the development of new processes, processing strategies to address the processing challenges generated by their complex structure.

3. Conclusions

Plants have always been an indispensable source for human life and evolution, providing not only essential nutrients but also a wide range of bioactive compounds with remarkable therapeutic value. The evaluation of the chemical composition of plants, such as the analysis of important classes including polyphenols, terpenoids, or alkaloids, has allowed the elucidation of the mechanisms by which these compounds contribute to maintaining health, the prevention of diseases, and the improvement of various chronic conditions. This chapter highlights the importance and usefulness of some plant species from the Cucurbitaceae family. The study of the chemical composition, as well as the therapeutic potential of four species from this plant family, and of the ways of integrating them into food products, supports the development of a balanced diet, rich in natural antioxidants and anti-inflammatory compounds, with protective effects against oxidative stress, and metabolic imbalances. Finally, recent research has demonstrated once again that plants have not only a fundamental nutritional role but also a true therapeutic potential, directly contributing to improving the quality of life and supporting health.

Acknowledgements

This work was partially supported by Internal Grant 2025 no. 7956/31.03.2025 of Department of Chemistry, Physics and Environment, Faculty of Sciences and Environment, "Dunărea de Jos" University of Galați, "Targeted therapeutic molecules, anti-inflammatory and antitumor, isolated from exotic plants acclimatized in Romania" and ADER grant 5.2.1. no 521/18.07.2023, "Conservation and valorization of the genetic heritage of aromatic and medicinal species that can be cultivated on the territory of Romania."

Author details

Anna Busuioc, Bianca Furdui, Andreea-Veronica Botezatu
and Rodica-Mihaela Dinică*
Department of Chemistry, Physics and Environment, Faculty of Science and Environment, "Dunărea de Jos" University of Galați, Galați, Romania

*Address all correspondence to: rodica.dinica@ugal.ro

References

[1] Cheng JT. Review: Drug therapy in Chinese traditional medicine. Journal of Clinical Pharmacology. 2000;**40**(5):445

[2] Van Raaij J, Hendriksen M, Verhagen H. Potential for improvement of population diet through reformulation of commonly eaten foods. Public Health Nutrition. 2009;**12**(3):325-330

[3] Xu X, Liu A, Hu S, Ares I, Martínez-Larrañaga MR, Wang X, et al. Synthetic phenolic antioxidants: Metabolism, hazards and mechanism of action. Food Chemistry. 2021;**353**:1-15

[4] Shahidi F. Antioxidants in food and food antioxidants. Nahrung – Food. 2000;**44**(3):158-163

[5] Lutz M. Healthy sustainable food patterns and systems: A planetary urgency. Medwave. 2021;**21**(7):e8436

[6] Larson G, Piperno DR, Allaby RG, Purugganan MD, Andersson L, Arroyo-Kalin M, et al. Current perspectives and the future of domestication studies. Proceedings of the National Academy of Sciences of the United States of America. 2014;**111**(17):6139-6146

[7] Totin E, Ayimasse AF, Roncoli C, Schreinemachers P, Ayenan MAT, Harris J. Role of social institutions in shaping vegetable consumption across Africa: A review. Global Food Security. 2024;**41**:100775. DOI: 10.1016/j.gfs.2024.100775

[8] Oloyede FM, Adebooye OC. Effect of season on growth, fruit yield and nutrient profile of two landraces of *Trichosanthes cucumerina* L. African Journal of Biotechnology. 2005;**4**(10):1040-1044

[9] Matsusaka Y, Kawabata J. Evaluation of antioxidant capacity of non-edible parts of some selected tropical fruits. Food Science and Technology Research. 2010;**16**(5):467-472

[10] Memariani Z, Gorji N, Moeini R, Farzaei MH. Traditional uses. In: Phytonutrients in Food: From Traditional to Rational Usage. Cambridge, United Kingdom: Elsevier Inc., Woodhead Publishing Series in Food Science, Technology and Nutrition; 2019. pp. 23-66. DOI: 10.1016/B978-0-12-815354-3.00004-6

[11] Chomicki G, Schaefer H, Renner SS. Origin and domestication of Cucurbitaceae crops: Insights from phylogenies, genomics and archaeology. The New Phytologist. 2020;**226**(5):1240-1255

[12] Rolnik A, Olas B. Vegetables from the Cucurbitaceae family and their products: Positive effect on human health. Nutrition. 2020;**78**:110788. DOI: 10.1016/j.nut.2020.110788

[13] Nagarajaiah SB, Prakash J. Chemical composition and bioactive potential of dehydrated peels of benincasa hispida, luffa acutangula, and sechium edule. Journal of Herbs Spices & Medicinal Plants. 2015;**21**(2):193-202

[14] Ugbaja CC, Fawibe OO, Oyelakin AS, Fadimu IO, Ajiboye AA, Agboola DA. Comparative phytochemical and nutritional composition of *Trichosanthes cucumerina* (L.) and some *Solanum lycopersicum* (L.) cultivars in Nigeria. American Journal of Plant Sciences. 2017;**08**(02):297-309

[15] Kathpalia R, Bhatla SC. Plant mineral nutrition. In: Plant Physiology, Development and Metabolism. Singapore: Springer Singapore; 2018. pp. 37-81

[16] Khan T, Ali M, Khan A, Nisar P, Jan SA, Afridi S, et al. Anticancer plants: A review of the active phytochemicals, applications in animal models, and regulatory aspects. Biomolecules. 2020;**10**(1):1-30

[17] Balandrin MF, Klocke JA, Wurtele ES, Bollinger WH. Natural plant chemicals: Sources of industrial and medicinal materials. Science (80-). 1985;**228**(4704):1154-1160

[18] Zaynab M, Fatima M, Sharif Y, Zafar MH, Ali H, Khan KA. Role of primary metabolites in plant defense against pathogens. Microbial Pathogenesis. 2019;**137**:103728

[19] Shahat AA, Hidayathulla S, Khan AA, Alanazi AM, Al OT, Alqahtani AS, et al. Acta Tropica phytochemical profiling, antioxidant and anticancer activities of *Gastrocotyle hispida* growing in Saudi Arabia. Acta Tropica. 2019;**191**:243-247. DOI: 10.1016/j.actatropica.2019.01.013

[20] Souto EB, Sampaio AC, Campos JR, Martins-Gomes C, Aires A, Silva AM. Chapter 2—Polyphenols for skin cancer: Chemical properties, structure-related mechanisms of action and new delivery systems. In: Atta-ur-Rahman, editor. Bioactive Natural Products, Studies in Natural Products Chemistry. Vol. 63. Amsterdam, Netherlands: Elsevier; 2019. pp. 21-42

[21] Hao DC, Gu X-J, Xiao PG. Chemotaxonomy. In: Medicinal Plants. Cambridge, United Kingdom: Woodhead Publishing Limited, Elsevier; 2015. pp. 1-48

[22] Miljkovic M. Carbohydrates. 1st ed. New York, NY: Springer New York; 2009. pp. 1-28

[23] Mohammad NA, Anwar F, Mehmood T, Hamid AA, Muhammad K, Saari N. Phenolic compounds, tocochromanols profile and antioxidant properties of winter melon [*Benincasa hispida* (Thunb.) Cogn.] seed oils. Journal of Food Measurement and Characterization. 2019;**13**(2):940-948. DOI: 10.1007/s11694-018-0008-x

[24] Mitchell JW, Marth PC. Growth regulators. Soil Science. 1948;**65**:205

[25] Greco M, Chiappetta A, Bruno L, Bitonti MB. In *Posidonia oceanica* cadmium induces changes in DNA methylation and chromatin patterning. Journal of Experimental Botany. 2012;**63**(2):695-709

[26] Trouvelot S, Héloir MC, Poinssot B, Gauthier A, Paris F, Guillier C, et al. Carbohydrates in plant immunity and plant protection: Roles and potential application as foliar sprays. Frontiers in Plant Science. 2014;**5**:1-14

[27] Sajid M, Stone SR, Kaur P. Recent advances in heterologous synthesis paving way for future green-modular bioindustries: A review with special reference to isoflavonoids. Frontiers in Bioengineering and Biotechnology. 2021;**9**:1-18

[28] Bukhari SA, Farah N, Mahmood S, Altaf J, Mustafa G. Effects of seed priming with zinc sulfate on nutritional enrichment and biochemical fingerprints of *Momordica charantia*. Journal of Food Quality. 2021;**2021**:1-13

[29] Okunowo W. Nutritional and chemical evaluation of *Momordica charantia*. Journal of Medicinal Plant Research. 2016;**4**:2189-2193

[30] Villarreal-La Torre VE, Guarniz WS, Silva-Correa C, Cruzado-Razco L, Siche R. Antimicrobial activity and chemical composition of Momordica Charantia: A review. Pharmacognosy Journal. 2020;**12**(1):213-222

[31] Tan HF, Gan CY. Polysaccharide with antioxidant, α-amylase inhibitory and ACE inhibitory activities from *Momordica charantia*. International Journal of Biological Macromolecules. 2016;**85**:487-496. DOI: 10.1016/j.ijbiomac.2016.01.023

[32] Ferrara L. A fruit to discover: *Cucumis metuliferus* E.Mey Ex Naudin(Kiwano). Clinical Nutrition Metabolism. 2018;**1**(2):1-2

[33] Jimam NS, Wannang NN, Omale S, Gotom B. Evaluation of the hypoglycemic activity of cucumis metuliferus (Cucurbitaceae) fruit pulp extract in normoglycemic and alloxan-induced hyperglycemic rats. Journal of Young Pharmacists. 2010;**2**(4):384-387. DOI: 10.4103/0975-1483.71633

[34] Rajasree RS, Sibi PI, Francis F, William H. Phytochemicals of cucurbitaceae family—A review. International Journal of Pharmacognosy and Phytochemical Research. 2016;**8**(1):113-123

[35] Marchei E, Pellegrini M, Pacifici R, Zuccaro P, Pichini S. A rapid and simple procedure for the determination of ephedrine alkaloids in dietary supplements by gas chromatography-mass spectrometry. Journal of Pharmaceutical and Biomedical Analysis. 2006;**41**(5):1633-1641

[36] Lovegrove A, Edwards CH, De Noni I, Patel H, El SN, Grassby T, et al. Role of polysaccharides in food, digestion, and health. Critical Reviews in Food Science and Nutrition. 2017;**57**(2):237-253

[37] Atanasov AG, Zotchev SB, Dirsch VM, Orhan IE, Banach M, Rollinger JM, et al. Natural products in drug discovery: Advances and opportunities. Nature Reviews. Drug Discovery. 2021;**20**(3):200-216

[38] Phillipson JD. Phytochemistry and medicinal plants. Phytochemistry. 2001;**56**(3):237-243

[39] Ma Y, Zhang L, Ma X, Bai K, Tian Z, Wang Z, et al. Saccharide mapping as an extraordinary method on characterization and identification of plant and fungi polysaccharides: A review. International Journal of Biological Macromolecules. 2024;**275**(P1):133350. DOI: 10.1016/j.ijbiomac.2024.133350

[40] Cao X, Du X, Jiao H, An Q, Chen R, Fang P, et al. Carbohydrate-based drugs launched during 2000-2021. Acta Pharmaceutica Sinica B. 2022;**12**(10):3783-3821

[41] Vázquez L, Corzo-Martínez M, Arranz-Martínez P, Barroso E, Reglero G, Torres C. Bioactive lipids. In: Reference Series in Phytochemistry. Cham, Switzerland: Springer Nature; 2019. pp. 467-527

[42] Teoh ES. Secondary metabolites of plants. In: Medicinal Orchids of Asia. Cham: Springer International Publishing; 2016. pp. 59-73

[43] Debjyoti P, Dey TK, Chakraborty A, Dhar P. Chapter 13—Promising functional lipids for therapeutic applications. In: Grumezescu AM, Holban AMBT, editors. Role of Materials Science in Food Bioengineering. Handbook of Food Bioengineering. London, United Kingdom: Academic Press, Elsevier; 2018. pp. 413-449

[44] Orhan N, Orhan IE, Ergun F. Insights into cholinesterase inhibitory and antioxidant activities of five Juniperus species. Food and Chemical Toxicology. 2011;**49**(9):2305-2312

[45] Ankita S, Parminder K, Ruby G. Phytochemical screening and

antimicrobial assay of various seeds extract of Cucurbitaceae family. International Journal of Applied Biology and Pharmaceutical Technology. 2012;**3**(3):401-409. Available from: www.ijabpt.com

[46] Lampi AM, Moreau RA, Piironen V, Hicks KB. Pearling barley and rye to produce phytosterol-rich fractions. Lipids. 2004;**39**(8):783-787

[47] Samad NB, Debnath T, Jin HL, Lee BR, Park PJ, Lee SY, et al. Antioxidant activity of *Benincasa hispida* seeds. Journal of Food Biochemistry. 2013;**37**(4):388-395

[48] Batmanabane V, Sivakumar R. Steroids in leptospiral uveitis: Does the route of administration matter? Journal of Pharmacology and Pharmacotherapeutics. 2012;**3**(1):58-60

[49] Liu HS, Wen LL, Chu PL, Lin CY. Association among total serum isomers of perfluorinated chemicals, glucose homeostasis, lipid profiles, serum protein and metabolic syndrome in adults: NHANES, 2013-2014. Environmental Pollution. 2018;**232**:73-79. DOI: 10.1016/j.envpol.2017.09.019

[50] Islam MT, Quispe C, El-Kersh DM, Shill MC, Bhardwaj K, Bhardwaj P, et al. A literature-based update on *Benincasa hispida* (Thunb.) Cogn.: Traditional uses, nutraceutical, and phytopharmacological profiles. Oxidative Medicine and Cellular Longevity. 2021;**2021**:1-19

[51] Ekeke C, Ogazie CA, Agbagwa IO. Anatomical and phytochemical studies on *Benincasa hispida* (Thunb.) Cogn. (Cucurbitaceae). Notulae Scientia Biologicae. 2019;**11**(1):102-111

[52] Amagon K. Flavonoids extracted from fruit pulp of *Cucumis metuliferus* have antiviral properties. British Journal of Pharmaceutical Research. 2012;**2**(4):249-258

[53] Graham RD, Welch RM, Bouis HE. Addressing micronutrient malnutrition through enhancing the nutritional quality of staple foods: Principles, perspectives and knowledge gaps. Advances in Agronomy. 2001;**70**:77-142

[54] Oyeleke GO, Adebisi GA. Evaluation of snake tomato (trichosanthes cucumerina) seed and extracted seed-oils for relevant applications. Innovative Science and Technology. 2022;**1**(2):14-22

[55] Machewad GM, Sawate AR, Zubair S, Kshirsagar RB. Studies on fatty acid profile, physico-chemical properties and phyto-esterol contents of bitter gourd (*Momordica charantia*) seed oil. The Pharma Innovation Journal. 2021;**10**(3):399-402

[56] Qadrie ZL, Hawisa NT, Khan MWA, Samuel M, Anandan R. Antinociceptive and anti-pyretic activity of *Benincasa hispida* (Thunb.) Cogn. in Wistar albino rats. Pakistan Journal of Pharmaceutical Sciences. 2009;**22**(3):287-290

[57] Omokhua-Uyi AG, Van Staden J. Phytomedicinal relevance of south African Cucurbitaceae species and their safety assessment: A review. Journal of Ethnopharmacology. 2020;**259**:112967. DOI: 10.1016/j.jep.2020.112967

[58] Busuioc AC, Botezatu AVD, Furdui B, Vinatoru C, Maggi F, Caprioli G, et al. Comparative study of the chemical compositions and antioxidant activities of fresh juices from romanian cucurbitaceae varieties. Molecules. 2020;**25**(22):1-17

[59] Šovljanski O, Šeregelj V, Pezo L, Šaponjac VT, Vulić J, Cvanić T, et al. Horned melon pulp, peel, and seed: New

insight into phytochemical and biological properties. Antioxidants. 2022;**11**(5):1-18

[60] Dey P, Kundu A, Kumar A, Gupta M, Lee BM, Bhakta T, et al. Analysis of alkaloids (indole alkaloids, isoquinoline alkaloids, tropane alkaloids). In: Recent Advances in Natural Products Analysis. Amsterdam, Netherlands: Elsevier; 2020. pp. 505-567

[61] Feng X, Cao S, Qiu F, Zhang B. Traditional application and modern pharmacological research of *Artemisia annua* L. Pharmacology & Therapeutics. 2020;**216**:107650. DOI: 10.1016/j.pharmthera.2020.107650

[62] Fuller G, Nes WD. Plant Lipids and Their Interactions. Washington, DC: ACS Symposium Series American Chemical Society; 1987. pp. 2-8

[63] Goggin FL, Shah J, Gillaspy G. Editorial: Lipid metabolism and membrane structure in plant biotic interactions. Frontiers in Plant Science. 2022;**13**:1-4

[64] Upadhyay A, Agrahari P, Singh DK. A review on salient pharmacological features of momordica charantia. International Journal of Pharmacology. 2015;**11**(5):405-413

[65] Jimam NS, Wannang NN, Anuka JA, Omale S, Falang K, Adolong AA. Histopathologic effects of *C. metuliferus* E Mey (Cucurbitaceae) fruits in albino rats. International Journal of Pharmaceutical Sciences and Research. 2011;**2**(8):2190-2194

[66] Alimi A, Idowu A, Akinde A, Fasina K, Taofik T, Aigbemekhe J. In vitro antibacterial potentials of sensitive plant (*Mimosa pudica*) and bitter lemon (*Momordica charantia*) leaf extracts and synthetic antibiotics against some bacteria isolates of *Clarias gariepinus*. Egyptian Journal of Agricultural Research. 2021;**99**(1):20-25

[67] Ezekaibeya AC, Nnenna AO, Kenechukwu OC. Proximate, phytochemical and vitamin compositions of *Cucumis metuliferus* (horned melon) rind. Journal of Complementary and Alternative Medical Research. 2020;**9**(3):40-50

[68] Usman A, Abdulrahman FI, Usman A. Qualitative phytochemical screening and in vitro antimicrobial effects of methanol stem bark extract of *Ficus thonningii* (Moraceae). African Journal of Traditional, Complementary, and Alternative Medicines. 2009;**6**(3):289-295

[69] Nandecha C, Nahata A, Dixit VK. Effect of *Benincasa hispida* fruits on testosterone-induced prostatic hypertrophy in albino rats. Current Therapeutic Research, Clinical and Experimental. 2010;**71**(5):331-343

[70] Okonwu K, Muonekwu JE. Potentials of underexploited seed of *Trichosanthes cucumerina* Linn. Journal of Applied Sciences and Environmental Management. 2019;**23**(5):791-797. PRINT ISSN: 1119-8362; Electronic ISSN: 1119-8362

[71] Rohajatien U, Harijono ET, Sriwahyuni E. Bioactive compound and nutritious characteristic of bitter melon fruit (*Momordica charantia* L.). Russian Journal of Agricultural and Socio-Economic Sciences. 2018;**79**(7):308-316

[72] Joseph B, Jini D. Antidiabetic effects of *Momordica charantia* (bitter melon) and its medicinal potency. Asian Pacific Journal of Tropical Disease. 2013;**3**(2):93-102

[73] Bortolotti M, Mercatelli D, Polito L. *Momordica charantia*, a nutraceutical

approach for inflammatory related diseases. Frontiers in Pharmacology. 2019;**10**:1-9

[74] Reyes BAS, Bautista ND, Tanquilut NC, Anunciado RV, Leung AB, Sanchez GC, et al. Anti-diabetic potentials of *Momordica charantia* and *Andrographis paniculata* and their effects on estrous cyclicity of alloxan-induced diabetic rats. Journal of Ethnopharmacology. 2006;**105**(1-2):196-200

[75] Sur S, Ray RB. Diverse roles of bitter melon (*Momordica charantia*) in prevention of oral cancer. Journal of Cancer Metastasis and Treatment. 2021;**7**:1-11

[76] Fajinmi OO, Olarewaju OO, Arthur GD, Naidoo K, Coopoosamy R. Cucurbitaceae species used as traditional medicine in West Africa. Journal of Medicinal Plants for Economic Development. 2022;**6**(1):1-9

[77] Perumal V, Khatib A, Uddin Ahmed Q, Fathamah Uzir B, Abas F, Murugesu S, et al. Antioxidants profile of *Momordica charantia* fruit extract analyzed using LC-MS-QTOF-based metabolomics. Food Chemistry: Molecular Sciences. 2021;**2**:100012

[78] Manimaran R, Murugu Mohan Kumar K. Experimental analysis on unmodified diesel engine characterization with novel biodiesel blends extracted from waste *Trichosanthes cucumerina* seeds. Energy Sources Part A: Recovery, Utilization, and Environmental Effects. 2021;**47**(1):1-21. DOI: 10.1080/15567036.2021.1905110

[79] Kumar V, Barwal A, Sharma N, Shafi D, Pradeep M, Vikas K. Therapeutic proteins: Developments, progress, challenges, and future perspectives. 3 Biotech. 2024;**14**:1-32. DOI: 10.1007/s13205-024-03958-z

[80] Ajuru M, Nmom F. A review on the economic uses of species of Cucurbitaceae and their sustainability in Nigeria. American Journal of Plant Biology. 2017;**2**(1):17-24. Available from: http://www.sciencepublishinggroup.com/j/ajpb

[81] Yang Y. Vitamins. In: Kan J, Chen K, editors. Essentials of Food Chemistry. Singapore: Springer Singapore; 2021. pp. 255-290

[82] Miret JA, Cela J, Bezerra LA, Arrom L, Juvany M, Müller M, et al. Application of a rapid and sensitive method for hormonal and vitamin E profiling reveals crucial regulatory mechanisms in flower senescence and fruit ripening. Journal of Plant Growth Regulation. 2014;**33**(1):34-43

[83] Al-snafi AE. Therapeutic properties of medicinal plants: A review of plants with cardiovascular effects (part 1). Pharmacology & Toxicology. 2015;**5**(3):163-176

[84] Mukherjee PK, Singha S, Kar A, Chanda J, Banerjee S, Dasgupta B, et al. Therapeutic importance of Cucurbitaceae: A medicinally important family. Journal of Ethnopharmacology. 2022;**282**:114599. DOI: 10.1016/j.jep.2021.114599

[85] Saeed F, Afzaal M, Niaz B, Arshad MU, Tufail T, Hussain MB, et al. Bitter melon (*Momordica charantia*): A natural healthy vegetable. International Journal of Food Properties. 2018;**21**(1):1270-1290

[86] Dhillon NPS, Lin CC, Sun Z, Hanson PM, Ledesma DR, Habicht SD, et al. Varietal and harvesting stage variation in the content of carotenoids, ascorbic acid and tocopherols in the fruit of bitter gourd (*Momordica charantia* L.). Plant Genetic Resources:

Characterisation and Utilisation. 2017;**15**(3):248-259

[87] Sung Goo K, Ashari S, Basuki N, Sugiharto AN. The bitter gourd *Momordica charantia* L.: Morphological aspects, charantin and vitamin C contents. IOSR Journal of Agriculture and Veterinary Science. 2016;**09**(10):76-81

[88] Al Snafi AE. The pharmacological importance of *Benincasa hispida*. A review. International Journal of Pharmaceutical Sciences and Research. 2013;**4**(12):165-170

[89] Oyesola O, Osonuga I, Taiwo O, Owoeye I. Leaf extract of *Momordica charantia*: Effects on liver enzymes, duodenum antioxidant enzymes, and disaccharidases with liver and duodenum histology. National Journal of Physiology, Pharmacy and Pharmacology. 2021;**11**(0):1

[90] Manimaran R, Murugu Mohan Kumar K, Sathiya Narayanan N. Synthesis of bio-oil from waste *Trichosanthes cucumerina* seeds: A substitute for conventional fuel. Scientific Reports. 2020;**10**(1):1-12. DOI: 10.1038/s41598-020-74130-9

[91] Patel RK, Patel SB, Shah JG. Anti-urolithiatic activity of ethanolic extract of seeds of *Benincasa hispida* (Thumb). Pharmacology. 2011;**3**:586-591

[92] Shetty BV, Arjuman A, Jorapur A, Samanth R, Yadav SK, Valliammai N, et al. Effect of extract of *Benincasa hispida* on oxidative stress in rats with indomethacin induced gastric ulcers. Indian Journal of Physiology and Pharmacology. 2008;**52**(2):178-182

[93] Schaefer H, Renner SS. Phylogenetic relationships in the order Cucurbitales and a new classification of the gourd family (Cucurbitaceae). Taxon. 2011;**60**(1):122-138

[94] Lay MM, Karsani SA, Mohajer S, Abd Malek SN. Phytochemical constituents, nutritional values, phenolics, flavonols, flavonoids, antioxidant and cytotoxicity studies on *Phaleria macrocarpa* (Scheff.) Boerl fruits. BMC Complementary and Alternative Medicine. 2014;**14**(1):1-12

[95] Harborne JB. Common fragrance and flavor materials. Phytochemistry. 1998;**48**:585

[96] Martiniakova M, Babikova M, Mondockova V, Blahova J, Kovacova V, Omelka R. The role of macronutrients, micronutrients and flavonoid polyphenols in the prevention and treatment of osteoporosis. Nutrients. 2022;**14**(3):523

[97] Bimakr M, Rahman RA, Ganjloo A, Taip FS, Adzahan NM, Sarker MZI. Characterization of valuable compounds from winter melon (*Benincasa hispida* (Thunb.) Cogn.) seeds using supercritical carbon dioxide extraction combined with pressure swing technique. Food and Bioprocess Technology. 2016;**9**(3):396-406

[98] Heimler D, Vignolini P, Dini MG, Vincieri FF, Romani A. Antiradical activity and polyphenol composition of local Brassicaceae edible varieties. Food Chemistry. 2006;**99**(3):464-469

[99] Fofana S, Gnoula C, Ouédraogo M, Palé E, Nébié RH, Nikiema JB, et al. DPPH radical scavenging and lipoxygenase inhibitory effects in extracts from *Erythrina senegalensis* (Fabaceae) DC. African Journal of Pharmacy and Pharmacology. 2016;**10**(11):185-191

[100] Belščak-Cvitanović A, Durgo K, Huđek A, Bačun-Družina V, Komes D. Overview of polyphenols and their properties. In: Polyphenols: Properties, Recovery, and Applications. Duxford,

United Kingdom: Woodhead Publishing an imprint of Elsevier; 2018. pp. 3-44

[101] Sabale V, Kunjwani H, Sabale P. Formulation and in vitro evaluation of the topical antiageing preparation of the fruit of *Benincasa hispida*. Journal of Ayurveda and Integrative Medicine. 2011;**2**(3):124-128

[102] Bessa C, Francisco T, Dias R, Mateus N, de Freitas V, Pérez-Gregorio R. Use of polyphenols as modulators of food allergies. From chemistry to biological implications. Frontiers in Sustainable Food Systems. 2021;**5**:1-18

[103] Shakya A, Chaudhary SK, Bhat HR, Gogoi N, Ghosh SK. A rapid high-performance thin-layer chromatographic method to estimate quercetin in benincasa hispida (Thunb.) Cogn. fruit pulp. Journal of Planar Chromatography - Modern TLC. 2019;**32**(6):467-474

[104] Fan M, Kim EK, Choi YJ, Tang Y, Moon SH. The role of momordica charantia in resisting obesity. International Journal of Environmental Research and Public Health. 2019;**16**(18):1-17

[105] Kenny O, Smyth TJ, Hewage CM, Brunton NP. Antioxidant properties and quantitative UPLC-MS analysis of phenolic compounds from extracts of fenugreek (*Trigonella foenum-graecum*) seeds and bitter melon (*Momordica charantia*) fruit. Food Chemistry. 2013;**141**(4):4295-4302

[106] Perez JL, Jayaprakasha GK, Patil BS. Metabolite profiling and in vitro biological activities of two commercial bitter melon (*Momordica charantia* Linn.) cultivars. Food Chemistry. 2019;**288**:178-186. DOI: 10.1016/j.foodchem.2019.02.120

[107] Thiruvengadam M, Praveen N, Maria John KM, Yang YS, Kim SH, Chung IM. Establishment of *Momordica charantia* hairy root cultures for the production of phenolic compounds and determination of their biological activities. Plant Cell, Tissue and Organ Culture. 2014;**118**(3):545-557

[108] Bukhari SA, Farah N, Mustafa G, Mahmood S, Naqvi SAR. Magneto-priming improved nutraceutical potential and antimicrobial activity of *Momordica charantia* L. without affecting nutritive value. Applied Biochemistry and Biotechnology. 2019;**188**(3):878-892

[109] Thomford KP, Thomford AK, Yorke J, Yeboah R, Appiah AA. *Momordica charantia* L. for hyperlipidaemia: A randomised controlled assessment of the Ghanaian herbal medicinal product MCP-1. Journal of Herbal Medicine. 2021;**28**:100453. DOI: 10.1016/j.hermed.2021.100453

[110] Arawwawala LDAM, Thabrew MI, Arambewela LSR, Fernando N, Guruge LD. Antibacterial activity of *Trichosanthes cucumerina* Linn. extracts. International Journal of Pharmaceutical & Biological Archive. 2011;**2**(2):808-812

[111] Razeek N, Sittampalam T, Kapilan R. Effect of water content on the germination characteristics of *Trichosanthes cucumerina*, *Momocardia charantia* and *Abelmoschus esculentus* seeds. Journal Of Applied Biology & Biotechnology. 2016;**4**(5):7-13

[112] Adeeyo OA, Ogundare O, Salawu EO, Saka WA, Adeleke GE, Onaolapo OJ, et al. Oral administration of aqueous extract of *Trichosanthes cucumerina* may prevent diabetic renal abnormalities. World Journal of Young Researchers. 2011;**1**:4-9

[113] Svobodova B, Barros L, Calhelha RC, Heleno S, Alves MJ, Walcott S, et al.

Bioactive properties and phenolic profile of *Momordica charantia* L. medicinal plant growing wild in Trinidad and Tobago. Industrial Crops and Products. 2017;**95**:365-373. DOI: 10.1016/j.indcrop.2016.10.046

[114] Guarniz WAS, Canuto KM, Ribeiro PRV, Dodou HV, Magalhaes KN, Sá KM, et al. *Momordica charantia* L. variety from northeastern Brazil: Analysis of antimicrobial activity and phytochemical components. Pharmacognosy Journal. 2019;**11**(6):1312-1324

[115] Wen JJ, Li MZ, Gao H, Hu JL, Nie QX, Chen HH, et al. Polysaccharides from fermented *Momordica charantia* L. with *Lactobacillus plantarum* NCU116 ameliorate metabolic disorders and gut microbiota change in obese rats. Food & Function. 2021;**12**(6):2617-2630

[116] Azeez MA, Morakinyo JA. Electrophoretic characterization of crude leaf proteins in Lycopersicon and Trichosanthes cultivars. African Journal of Biotechnology. 2004;**3**(11):585-587

[117] Mitra S, Prova SR, Sultana SA, Das R, Nainu F, Bin ET, et al. Therapeutic potential of indole alkaloids in respiratory diseases: A comprehensive review. Phytomedicine. 2021;**90**:153649. DOI: 10.1016/j.phymed.2021.153649

[118] Ahamad J, Amin S, R. Mir S. *Momordica charantia* Linn. (Cucurbitaceae): Review on phytochemistry and pharmacology. Research Journal of Phytochemistry. 2017;**11**(2):53-65

[119] Ambadiang MMM, Atontsa BCK, Tankeo SB, Nayim P, Wamba BEN, Bitchagno GTM, et al. Bark extract of *Cassia sieberiana* DC. (Caesalpiniaceae) displayed good antibacterial activity against MDR gram-negative phenotypes in the presence of phenylalanine-arginine β-naphthylamide. BMC Complementary Medicine and Therapies. 2020;**20**(1):1-11

[120] Šeregelj V, Šovljanski O, Šaponjac VT, Vulić J, Ćetković G, Markov S, et al. Horned Melon (Cucumis metuliferus E. Meyer Ex. Naudin)—Current knowledge on its phytochemicals, biological benefits, and potential applications. Processes. 2022;**10**(1):1-14

[121] Evidente A, Kornienko A, Lefranc F, Cimmino A, Dasari R, Evidente M, et al. Sesterterpenoids with Anticancer Activity. Current Medicinal Chemistry. 2015;**22**(30):3502-3522

[122] Li W, Lin Z, Yang C, Wang Y, Qiao Y. Study on the chemical constituents of *Momordica charantia* L. leaves and method for their quantitative determination. Biomedical Research. 2015;**26**(3):415-419

[123] Aliero AA, Gumi AM. Studies on the germination, chemical composition and antimicrobial properties of *Cucumis metuliferus*. Annals of Biological Research. 2012;**3**(8):4059-4064. Available from: http://scholarsresearchlibrary.com/ABR-vol3-iss8/ABR-2012-3-8-4059-4064.pdf

[124] Ye DY, Qi YH, Cao SF, Wei BQ, Zhang HS. Histopathology combined with transcriptome analyses reveals the mechanism of resistance to *Meloidogyne incognita* in *Cucumis metuliferus*. Journal of Plant Physiology. 2017;**212**:115-124. DOI: 10.1016/j.jplph.2017.02.002

[125] Muribeca A d JB, PWP G, Paes SS, APA d C, PWP G, Viana J d S, et al. Antibacterial activity from *Momordica charantia* L. leaves and flavones enriched phase. Pharmaceutics. 2022;**14**(9):1-16

[126] Md, Setamam N, Jaafar Sidik N, Abdul Rahman Z, Che Mohd Zain CR. Induction of hairy roots by various strains of *Agrobacterium rhizogenes* in

different types of Capsicum species explants. BMC Research Notes. 2014;**7**(1):1-8

[127] Bai R, Yao C, Zhong Z, Ge J, Bai Z, Ye X, et al. Discovery of natural anti-inflammatory alkaloids: Potential leads for the drug discovery for the treatment of inflammation. European Journal of Medicinal Chemistry. 2021;**213**:113165

[128] Busuioc AC, Costea GV, Veronica A, Botezatu D, Furdui B, Dinica RM. *Cucumis metuliferus* L. fruits extract with antioxidant, anti-inflammatory, and antidiabetic properties as source of ursolic acid. Separations. 2023;**10**(5):274, 1-15

[129] Annapoorani CA, Manimegalai K. Evaluation of biological activity and qualitative analysis of 2, 5-dihydroxybenzoic acid from *Momordica charantia* fruit. International Journal of Pharmaceutical Sciences Review and Research. 2013;**22**(2):89-95

[130] Alshehri A, Ahmad A, Tiwari RK, Ahmad I, Alkhathami AG, Alshahrani MY, et al. In vitro evaluation of antioxidant, anticancer, and anti-inflammatory activities of ethanolic leaf extract of *Adenium obesum*. Frontiers in Pharmacology. 2022;**13**:1-12

[131] Arawwawala M, Thabrew I, Arambewela L. In vitro and in vivo evaluation of antioxidant activity of *Trichosanthes cucumerina* aerial parts. Acta Biologica Hungarica. 2011;**62**(3):235-243

[132] Patel S, Rauf A. Edible seeds from Cucurbitaceae family as potential functional foods: Immense promises, few concerns. Biomedicine & Pharmacotherapy. 2017;**91**:330-337. DOI: 10.1016/j.biopha.2017.04.090

[133] Shivanagoudra SR, Perera WH, Perez JL, Athrey G, Sun Y, Wu CS, et al. In vitro and in silico elucidation of antidiabetic and anti-inflammatory activities of bioactive compounds from *Momordica charantia* L. Bioorganic and Medicinal Chemistry. 2019;**27**(14):3097-3109. DOI: 10.1016/j.bmc.2019.05.035